Bigfoot Encounters in Ohio

George Clappison (left) and Joedy Cook of the Ohio Bigfoot Research & Study Group are seen in this photograph holding casts of footprints found in Ohio. These two researchers were highly involved in bigfoot research in Ohio. They have personally investigated numerous sightings, and their group became a primary source for both current and historical information on the Ohio bigfoot phenomenon.

BIGFOOT
Encounters in Ohio
Quest for the Grassman

Christopher L. Murphy
in association with
Joedy Cook and George Clappison

ISBN-13: 978-0-88839-157-5

Copyright © 2006 Christopher L. Murphy

2018 Reprint

Cataloging in Publication Data

Murphy, Christopher L. (Christopher Leo), 1941–
    Bigfoot Encounters in Ohio : quest for the grassman / Christopher L. Murphy in association with Joedy Cook and George Clappison.

Includes bibliographical references and index.
ISBN 0-88839-607-4

    1. Sasquatch—Ohio. I. Cook, Joedy II. Clappison, George III. Title.
QL89.2.S2M867 2005   001.944   C2005-905732-7

All rights reserved. No part of this publication may be reproduced, stored in a retrieval system or transmitted, in any form or by any means, electronic, mechanical, photocopying, recording, or otherwise, without the prior written permission of Hancock House Publishers.

Printed in he USA

Cover Illustration: Christopher L. Murphy

We acknowledge the financial support of the Government of Canada through the Book Publishing Industry Development Program (BPIDP) for our publishing activities.

Crypto Editions is an imprint of Hancock House Publishers

Published simultaneously in Canada and the United States by

HANCOCK HOUSE PUBLISHERS LTD.
19313 Zero Avenue, Surrey, B.C. Canada V3Z 9R9
(604) 538-1114 Fax (604) 538-2262

HANCOCK HOUSE PUBLISHERS
#104-4550 Birch Bay-Lynden Rd, Blaine, WA U.S.A. 98230-9436 (800) 938-1114 Fax (800) 983-2262

Website: www.hancockhouse.com
Email: sales@hancockhouse.com

CONTENTS

	Acknowledgments	6
	Introduction	7
1.	**Ohio's Profile**	11
	Geography	11
	Bigfoot's Possible Migration Route	12
	Bigfoot Origination Theory	12
	Early Ohio Dwellers	15
	Ohio's Bigfoot Credibility	18
2.	**Bigfoot "Recognition"**	25
3.	**Bigfoot Speculations**	33
4.	**Ohio Bigfoot in Review**	39
5.	**Sighting Considerations**	97
6.	**Ohio Black Bears and Bigfoot**	100
7.	**Ohio Deer Kills and Bigfoot**	102
8.	**The Patterson Legacy**	111
9.	**Intriguing Questions**	117
10.	**Where to From Here?**	119
11.	**Mysterious Tracings**	121
	Krao	121
	Julia Pastrana	122
	Zana	124
	The Karapetian Hominid	127
	Bassou	128
	Chinese Bigfoot Cross-breed	129
	Francis de Loys' Man-beast	130
	The Yeti	131
	Appendix	134
	Ohio Bigfoot Incidents – County Order	134
	Ohio Bigfoot Incidents – Date Order	138
	Ohio's Counties	141
	Bibliography	142
	Photograph/Illustration Sources and Copyrights	143
	General Index	146

Acknowledgments

Special thanks are extended to the following researchers and others who have been instrumental in the compilation of this work through their own investigations or by providing information, photographs, artwork, and advice:

Reverend Lee Birt
Chris Coffey
Dean Cottrill
Erik Dahinden
Martin Dahinden
René Dahinden
Terry Endres
Henry Franzoni
John Green
Robert Gardiner
Don Keating
Roger Knights
Yvon Leclerc
Dan Murphy
Matt Moneymaker
Robert W. Morgan
John S. Sawvel Jr.
Ron Schaffner
Tom Steenburg
Kenny Young
and
The many witnesses who have come forward and shared their experiences.

NOTE: This acknowledgment does not imply endorsement of this book by the individuals shown. It merely indicates appreciation for the work they have done in the field of bigfoot research.

Introduction

Since ancient times, mankind has been confronted with various phenomena and natural mysteries. Many areas of North America have recurring incidents that have baffled us since recorded history.

The bigfoot/sasquatch mystery has existed for centuries. Many cultures around the globe have legends of hairy hominids. The more persistent reports originate from the Pacific Northwest regions of the United States and Canada. The media has played an important role in distributing reports. Newspapers such as the *San Francisco Chronicle,* the *New York Globe* and many other major newspapers have published articles that described man-like creatures from alleged eyewitness testimony. There have also been numerous television documentaries in which bigfoot researchers and scientists have been interviewed on the subject.

The "bigfoot" we know today achieved prominence in 1958. In that year, a road construction worker reported to the press large human-like footprints he had found in the Bluff Creek, California area. The story was carried nation-wide and "bigfoot," as it were, stepped into the limelight. Nine years later, what is believed to be a bigfoot was captured on film, also in the Bluff Creek, California area, by Roger Patterson and Bob Gimlin.

Unfortunately, the nature of this creature has led to a lot of tabloid miss-information, wild speculation, hoaxing and general ridicule. As a result, many people who have sightings are reluctant to come forward with information. More importantly, major research organizations and most scientists have adopted a "hands off" policy to avoid risking their credibility.

The sheer number of reported bigfoot incidents (sightings, footprint findings) in North America, coupled with the tremendous range of such evidence, is far too great to dismiss as the work of hoaxers.

BIGFOOT INCIDENTS REPORTED OVER ABOUT 100 YEARS

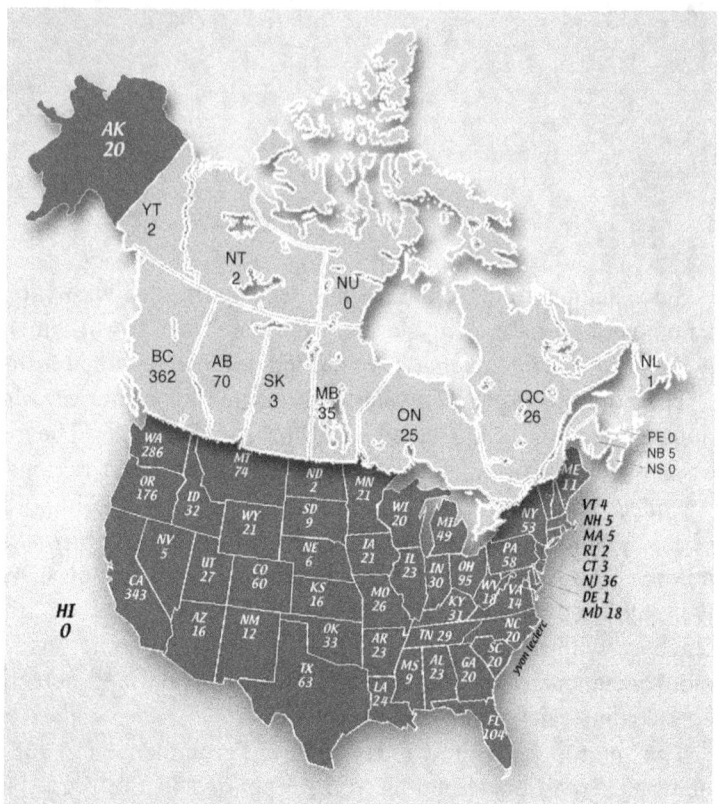

The total number of incidents shown is 2,557. The breakdown is based on a review of available statistics on plotted or "generally accepted" incidents as of the end of 2003. Numerous unpublished reports, however, are not included in the figures shown. It will be noted that in this work over 150 *possible* incidents are referenced for Ohio.

With human population increases and improved communications, the number of *reported* possible sasquatch incidents has significantly increased, and they are said to be around 400 incidents per year. It is reasoned that the number of incidents reported may be just a fraction of the actual number because many people are reluctant to report their experiences. The above map showing reports on record provides an insight into the extent of the phenomenon. Certainly, our relentless "invasion" into wilderness regions plays a major role in the figures shown.

While the sasquatch is not yet recognized by the scientific community, it has caught the attention of many prominent anthropologists and other professionals. Some of these people have spent a great deal of time, energy and resources in attempting to prove the creature's existence.

This book documents and discusses bigfoot related reports that originated within the state of Ohio, which ranks fifth in the number of reported bigfoot incidents in the United States, and sixth for North America. I have then reported the incidents by individual county. Certainly, bigfoot does not pay any attention to political boundaries, so such are really unrelated in a study of this nature. Nevertheless, they do provide a convenient method for presenting information and bring the subject "close to home," as it were.

Also included in this work is some general information about the bigfoot phenomenon on the Pacific coast (both in the United States and Canada). The purpose for including this information is to provide readers with a reasonably complete understanding of the subject.

Finally, for general interest, I have included a chapter on special people in history who were said to have ape-like characteristics, and hominids from other parts of the world.

It is hoped that this work will stimulate more interest in bigfoot research in Ohio.

About the Title
Use of the Ohio term "Grassman" to identify an unusual ape-man creature appears to go back to at least the turn of the last century. Apparently, sightings of the creature in tall grass (including the young of the species) on Ohio's plains resulted in the name. As the creature was somewhat terrifying in appearance, it appears the term was used in a foreboding sense with children (i.e., the *Grassman* will get you!). Descriptions of the Grassman are identical to those of bigfoot or the sasquatch; however, the Grassman appears to have some different habits or ways of life.

Note on Terminology
In this work I have chosen not to capitalize the words "bigfoot" or "sasquatch" but have left the words as shown in quoted material. Further, I have chosen to consider both words singular or plural in form.

OHIO'S TOPOGRAPHY

Chapter 1

Ohio's Profile

GEOGRAPHY: The land regions that constitute the state of Ohio were formed as the result of several glaciers moving down from the north thousands of years ago. These glaciers covered all but the southeast section of the state. The work of the glaciers produced four well-defined land regions, as follows: (1) the Great Lakes Plains, (2) the Till Plains, (3) the Bluegrass Region, and (4) the Appalachian Plateau.

The Great Lakes Plains are a narrow strip of land that borders Lake Erie. The entire area is primarily flat, save a few sandy ridges. These ridges once formed the rim of an ancient lake. The Till Plains stretch across central and western Ohio. Most of this region is rolling land with a few low hills. The Bluegrass Region is the smallest of Ohio's land regions. It lies wedged between the Till Plains and the Appalachian Plateau in the center of Ohio's southern boundary. It is bordered on the south by the Ohio River. The land surface is rolling. The Appalachian Plateau (also called the Allegheny Plateau) makes up the entire eastern section of Ohio. The northern third of this region is made up of fertile hills and valleys. The remaining two-thirds, which were not affected by the glaciers, has the most rugged terrain in the state.

The Ohio River stretches more than 450 miles (724km) along Ohio's eastern and southern border (Appalachian Region). The river winds through a narrow valley (the Ohio Valley) that is never more than two miles (3.22km) wide. The final formation of the Ohio Valley occurred about 18,000 years ago, during the last glacier period. About 4,000 years later, a warming trend occurred. The action of flowing water on the abundance of limestone produced the many caverns found throughout Ohio and Kentucky.

Indian legends tell us of many tunnel systems in the Ohio Valley region that are both natural and man-made. Indeed, ancient tunnels have been found that appear to have been made by early native people who came to this region.

BIGFOOT'S POSSIBLE MIGRATION ROUTE: If bigfoot do exist in North America, they have been here a very long time. We can reason that the creatures came here from Eurasia by crossing the land bridge that connected Eurasia to North America (the present Bering Strait). This passage was usable for at least 20,000 years and indeed was used both ways by human hunters on the trail of arctic game. However, by about 8000 B.C. anyone or anything that was in North America was here to stay if they did not possess a boat. By this time, melting ice sheets had drastically raised the sea level so the BeringStrait area could no longer be crossed on foot.

Given the 8000 B.C. "no return" time frame, we can say that

- Ice-limit 20,000 years ago.
- Land bridge caused by lower sea levels 20,000 years ago.
- Route of transmigration.

bigfoot have been here for *at least* 10,000 years. As to the maximum time, it is probably around 30,000 years, given the land bridge existed for *about* 20,000 years.

BIGFOOT ORIGINATION THEORY: If bigfoot do indeed exist, the main question to be answered is, what kind of creature is it? Certainly, the only way this question can be properly answered is by having an actual body of the creature (or body part), or at the very least, bones. Despite a few alleged killings of bigfoot, and at least one report of a rotting carcass, we still don't have any evidence of this nature.

Nevertheless, given the evidence we do have, considerable speculation has been made as to the creature's true identity. The most popular theory is that bigfoot belongs to a species called *Gigantopithecus blacki* that was assumed to have become extinct about 300,000 years ago. Evidence of this creature's existence is based on jawbones and teeth found in China and India. *Gigantopithecus blacki* is the largest primate that has ever been known to exist, and as such becomes a reasonable candidate for bigfoot.

Dr. Grover Krantz and model of a *Gigantopithecus blacki* created by William Munns and Russell Ciochon.

It is speculated that some of these ancient creatures found their way over the land bridge that once connected Eurasia with North America (previously discussed). In their new domain, these prehistoric immigrants apparently flourished and were not affected by the conditions that caused the extinction of their relatives who remained in Eurasia. Dr. Grover S. Krantz was the main proponent of the *Gigantopithecus blacki* theory. Based on a lower jawbone, Dr. Krantz constructed the entire skull of a *Gigantopithecus blacki* which is seen in the following photograph compared with a gorilla skull and human skull.

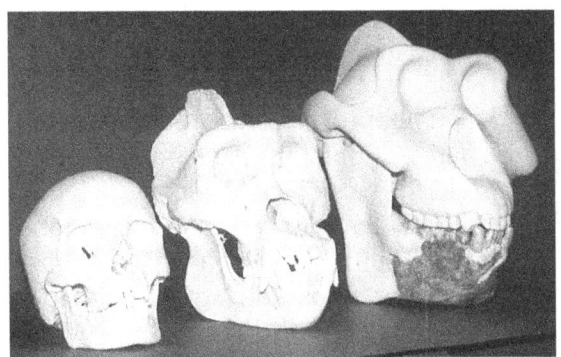

Human   Gorilla   *Gigantopithecus blacki*

THE BIRTH OF THE NAME "BIGFOOT"

A road was constructed into the Bluff Creek, California region in 1957, opening the area, which up to that time had been remote wilderness. Over the years, people in the region had noticed large human-like footprints and called whatever was making the prints "bigfoot." While at least one report had been provided to a local newspaper, no attention was paid to the matter. On August 27, 1958, Gerald (Jerry) Crew a road construction worker, saw such prints circling his parked bulldozer. Crew had heard of similar findings by a road gang about one year earlier at a location eight miles (12.9km) north. He showed the prints to his fellow workers, some of whom said they had also seen prints in the area. Crew saw additional prints about one month later and more on October 2, 1958. This time, he made a plaster cast of one of the prints. He took the cast to the *Humboldt Times* newspaper and related the story of his find. An *Associated Press* release (October 6, 1958) showed a photograph of Crew holding the cast. The release used the term "bigfoot," which resulted in this name becoming the recognized name for the creature in the United States.

Together with the footprints found at that time, there were also alleged sightings and other unusual incidents in the area. In later years, investigations revealed tracks of six different sizes, indicating that a number of bigfoot frequented the area. Footprint sizes ranged from 12.25 inches to 17 inches (31.1cm to 43.2cm) long. These facts made the Bluff Creek area a prime location for a possible bigfoot sighting.

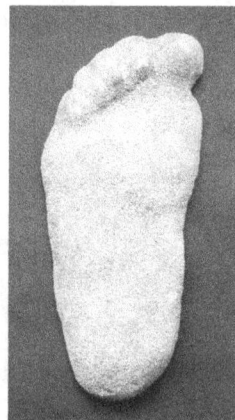

Copy of the cast made by Crew.

EARLY OHIO DWELLERS: In America's pre-historic times, it is said that a band of renegade giants, probably survivors of the Olmed and Toltec civilizations of Mexico, settled for a short time in the Ohio Valley, opening trade routes. Who were these giant people, and is it possible they are in some way connected with the bigfoot phenomenon? This question remains unanswered. However, we can reason that if there were giant native people in North America, they would have had giant ancestors. It may be from this ancestral line that a bigfoot "native" somehow became isolated or segregated and carried on to the present day.

This line of thinking probably originated after sasquatch or bigfoot were reported seen in British Columbia, Canada. The *Encyclopedia Canadiana* (1970) states in its entry on "Sasquatch" that, "The finding in 1932 of the remains of a long-extinct race of giants in Mexico gave some impetus to the belief that the remnants of a prehistoric race of troglodytes (i.e., the Sasquatch) may have survived in British Columbia." (No further information has surfaced on the "remains" to my knowledge.)

There appears to be some evidence supporting the fact that early "giants" lived in Ohio and its bordering states. In 1833, it was reported that a coal miner in Ohio broke into an ancient tunnel and discovered 17 fossilized human bodies that were very large. Also, 12 foot (3.7m) long human skeletons are said to have been found beneath the city of Lexington, Kentucky. Further, in the region surrounding Bowling Green, Ohio, very large human skeletons were reported found in the First Nations mounds in this area. Some skeletons were as large as eight feet (2.44m) in length.

Moreover, in 1876 two men exploring a cave in Louisville, Kentucky reported that they found a large underground room containing a stone vault. They opened the vault and found three skeletons nearly nine feet (2.74m) in length. Finally, in 1884 Indian mounds excavated in Christian County, Kentucky are said to have contained perfect skeletal remains of a large race of people.

Remarkably, the entire skeletons of a giant woman and her baby were found in Yosemite Valley, California in 1895. Here I will draw upon the work of Richard Smedley who provided the story in *Probe the Unknown* magazine, March 1975.

We are told that a party of miners in the valley noticed a pile of stones placed against the wall of a cliff. As the stones did not appear to be naturally placed, the miners removed the pile. Behind the stones they found a wall that had most definitely been made with knowledge of masonry. The joints between the rocks were all a uniform 1/8 inch (3mm) in thickness, and the men remarked that the wall was indeed a beautiful piece of stonework.

Mummified remains of a giant woman holding a child found in Yosemite Valley, California in 1895.

Thinking they had stumbled upon a lost treasure vault, the miners commenced immediately to tear down the wall. There was indeed a vault carved into the rock directly behind the wall. The vault measured 9.25 feet (2.82m) high, 18.5 feet (5.64m) deep and 8.33 (2.53m) feet wide. However, its only contents was what they deemed to be a single wrapped mummified corpse resting on a carved ledge. It was evident the vault had been created to serve as a tomb.

The mummy was very long (tall) measuring about 6 feet, 8 inches (2m). It was wrapped in what appeared to be animal skins and covered with a layer of fine gray powder. The miners removed some of the animal skins around the upper part of the mummy and to their surprise found the remains of a woman holding a child.

When the relic was placed before men of science, all agreed that the height of the woman in life would have been about 7 feet. As women are generally shorter than men, it was reasoned that males of the same "species" would have generally been 8 feet tall or taller.

All of the scientists further agreed that the mummies predated the Christian era. This fact ruled out ancestors of local Native people, the Ahwahneechees, who it is believed settled in the area no more than 1,000 years ago. Further, the Ahwahneechees were small people, generally being around 5 feet, 3 inches in height.

It is interesting to note, however, that Native folklore tells of a terrifying giant who came to the valley long before the white man arrived. The giant's name was Oo-el-en and he captured and ate native people. The Ahwahneechees eventually overpowered, killed and burned the menace, thereby ruling out any connection with the giant woman found in the vault.

It would have been interesting to know what the native people thought of the mummies. Unfortunately, it is too late now to do any research in this connection as the last of the Ahwahneechee people died in 1946.

All the evidence supporting these findings has since been lost in history, or perhaps it is locked away in various museum basements, waiting to be rediscovered. However, there is in existence a remarkable giant human skull found in Nevada in 1924 that might provide some further tangible evidence supporting North American giants.

The skull was found in the Lovelock Cave area. Other bones from this area indicated they belonged to a race of people ranging from 6.5 feet to 10 feet (2m to 3m) in height. These measurements were based on the skeleton femurs. However, anthropologists disagreed with the calculations, stating that the tallest they could justify was 5 feet, 11 inches (1.8m). Nevertheless, the skull remains a mystery.

On the other side of the coin, there is also possible evidence that Ohio was the home of a race of pygmy natives. During the last century, Dr. S. P. Hildreth, a professional archaeologist and native of Marietta, Ohio, investigated many aboriginal remains in the Ohio Adena mound-builder area. This area is located one mile south of the city of Coshocton, on a bluff overlooking the Killbuck River. In 1853, Hildreth found what may have been a pygmy cemetery. The cemetery had more than three thousand stone-carved graves, each containing a small human skeleton. The skeletons ranged in height from 3 to 4.5 feet (91cm to 135cm). Wood was discovered around the bones, indicating that wooden coffins were used. Archaeologists remain divided in their opinions on this discovery. Many believe the remains were those of children, but others maintain the crania of the individual skeletons indicates a human adult.

While pygmy populations would not directly support the possibility of bigfoot creatures in Ohio, there is an indirect link. It is apparent the Ohio region has attracted a diverse assortment of humanity and other animals. Can it not be reasoned that bigfoot was included?

OHIO'S "BIGFOOT" CREDIBILITY: The first question that presents itself when one thinks about bigfoot and the state of Ohio, is the ability of this state to support a creature of bigfoot's nature. The creature, as we believe it to be, is a very large mammal that mainly sustains itself on uncultivated vegetation, fish and small mammals. These factors demand fairly vast regions of wilderness that have an abundance of water. Early bigfoot researchers realized this condition and were quick to point out that the Pacific Northwest was probably the last stronghold for the creature, if indeed it did exist. Certainly, the Pacific Northwest has the largest wilderness areas; however, Ohio, together with many other states, still has significant "pockets" of wilderness. Moreover, it is noted that of Ohio's 41,000 square-mile (56,800km^2) area, about 81% is rural land or forest areas. The official breakdown (1992) is as follows: cropland 47%; forest 24%; range 10%; urban 8%, other 10%, federal 1%.

When we look at the geography of Ohio, we see that a great water resource, the Ohio River, stretches from southeast Pennsylvania to Paducah, Kentucky where it empties into the Mississippi River. The Ohio River, with its numerous tributaries, flows through national al forests and state parks with thousands of acres of wilderness and mountainous terrain. Ohio's Wayne National Forest, for example, comprises 108,822 acres. Further, Ohio's 17 state parks comprise over 150,000 acres. In bordering states, there is the Daniel Boone National Forest (Kentucky), with over 670,000 acres, and the Hoosier National Forest (Indiana), with 120,381 acres. These forests, according to the United States Department of the Interior, consist mainly of elm, ash, cottonwood, and pine trees. They are thick, dense forests that support an abundance of wildlife. The amount of rainfall required to maintain these forests is considerable and in some areas parallels that of the Pacific Northwest. Here we might mention that bigfoot would hold no allegiance to state lines.

Of all the regions in Ohio, the Ohio Valley appears to be the most likely habitat for bigfoot. Natural caves would provide protection, and there is an abundance of water together with dense forests for food sources.

> We learn that during the Pleistocene Epoch, animals reached great proportions, and when driven from their natural habitats by the advance of ice, they proceeded towards the Ohio Valley. Their craving for salt led many into an area that is now known as Big Bone Lick State Park in Northern Kentucky. This area has many sulfur springs with large salt deposits.

Given these facts, it is highly possible Ohio could support a bigfoot population. Certainly, this population would not be as large as that in the Pacific Northwest. It is natural that larger wilderness regions would host higher populations – this applies to all animals.

Remarkably, long before possible bigfoot or sasquatch beings became common knowledge, First Nations people carved masks that appear to show ape-like creatures, as seen in the adjacent drawings. The masks depicted were found in the early 1900s. I believe they are considerably older. As there are no apes in North America (other than possible bigfoot creatures), we must wonder what inspired these people to make the carvings – were these the images of early giants? I will mention here that I was informed by a native elder that the first mask shown reflects a sacred belief in "mountain monkeys."

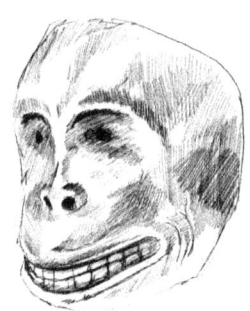

Nisga'a/Tsimshian masks showing ape-like features. (Drawings by Pete Travers.)

It has been suggested, however, that some natives may have seen a monkey brought to North America by a sailor. Intrigued with the creature, the native people used its features for their masks, and the designs were copied by various tribes. We might also reason that very early travelers who migrated into South America may have come back into North America with pet monkeys.

Just when First Nations people started to intentionally depict bigfoot or bigfoot-like creatures in their art is not certain. However, at this time there are numerous renderings. Perhaps the most revealing is the mask shown below that was made in 1938 by Ambrose Point of the Chehalis First Nations people in British Columbia, Canada.

This mask was created long before the Patterson/Gimlin film provided a reasonable description of the creature's facial features. However, the mask has a number

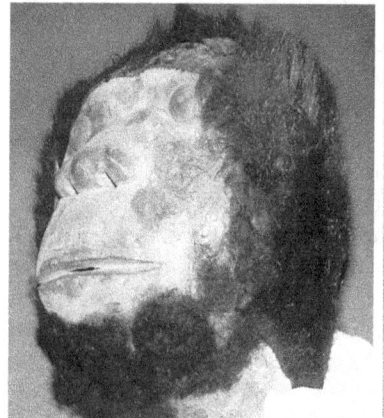
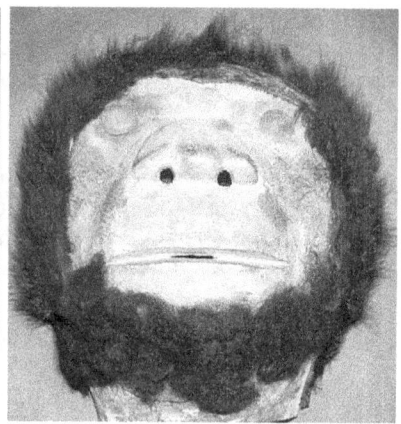

Sasquatch mask created by Ambrose Point, Chehalis First Nations, 1938.

of characteristics that match the creature in the film. We do not know if Ambrose Point actually saw a sasquatch, but such creatures are an integral part of Chehalis history. One might note that this mask and the other masks shown are not painted.

One of the earliest Ohio references to bigfoot-like creatures is in a book entitled, *A Buckeye Boyhood,* by William Venable (originally published in 1877 by the Robert Clark Co., Cincinnati). The book describes events during the period from 1836 to 1858. It relates the following:

> Tip heard a graybeard farmer tell, with serious air and bated breath, that a 'Bosjesman' had long been seen clambering

and crouching among the branches of a shell-bark hickory in the loneliest part of the swamp-woods on the road to Utica. What a 'Bosjesman' might be or do, no one could tell, but Tip conjectured that in all probability it was something between a gorilla and Sinbad's Old Man of the Mountain, and that it was particularly fond of the taste of cowardly blood. Soon after hearing the farmer's hoaxing story, the boy was sent on an errand to Utica, and as he rode by the swamp woods he felt very much as Tam O'Shanter must have felt,

Whiles glow'ring round wi' prudent cares,
Lest bogles catch him unawares.

The Delaware First Nations people of Northern Ohio certainly had a fear of some sort of "wild man." These people actually had a warning sign that they placed at the perimeter of "wild man country" to warn their people not to go into the region. The sign was a post with a carving that was dug into the ground. It extended about 10 feet (3.1m) above the ground. They also had a special ax for killing the creatures. While things of this nature are generally associated with myths, we are led to wonder – did the myths create the creature, or did the creature create the myths?

 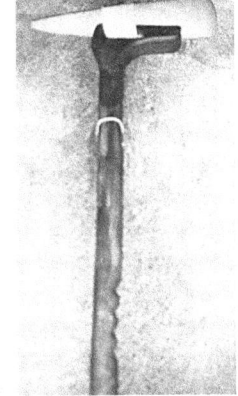

Left, Delaware First Nations "wild man" warning post. Right, ax used for killing "wild men."

NISGA'A LISIMS GOVERNMENT
COUNCIL OF ELDERS

During my research for a sasquatch exhibit at the Vancouver Museum in 2004, three council elders of the Nisga'a Lisims First Nations Government (British Columbia) shared their individual stories about the sasquatch in a letter to the museum. The following are their submissions, directly as recorded by Allison Nyce, manager, Ayuukhl Nisga'a Department. I have left the stories verbatim to preserve their integrity and unique presentation.

Horace Stevens: I remember when all the villages used to have public works and when they were talking about the big hairy man. The young men used to go hunting up on the mountain, when they would head out, they did not go for only one afternoon. They would go for a week and before they headed out to their hunting grounds, they would say where the canoes would land. They would stop at Nass Harbor and walk up the mountain until they came upon Ginluuak. They would go up a certain trail and come back down this way again, that is where they would put down what they were carrying. My father, Albert Stevens, was not a hunter; he told me that when he heard that the hunters were going out hunting he would grab his packsaddle. There would be about 8 or 10 of them but they were not all hunters, They would have those who would be the packers for them. They got the wool from the mountain goats; they know how to make the wool. They were very proud of themselves for making wool. They would make woolen socks, woolen pants and shirts so that a hunter would not catch any sickness while out hunting. My father did not tell *adaawak (traditional stories)* but he told us what they did back in those days. He told about the hunting expedition up on the mountain one time during the winter. They seen footprints in the snow at the shady area, they called it the footprints of the hairy man. One time they were looking down and seen a man walking, he would walk way down. They would always spot him at Nass Harbor. One time they went out hunting for about a day and a half just to get enough food and that is when they seen it again. While they were walking, they seen his footprints going across, it used to snow early a long time ago, late September or early October.

The little ponds used to be frozen over and they seen the footprints sunken in about one inch deep; they knew it was the hairy man. They called it the na<u>x</u>no<u>k</u> *(supernatural)*, because when they spotted him, he would disappear. I believe this story is true, no one has captured the hairy man; they have only seen him going by. We have also read this in the newspaper of other native people seeing the Sasquatch, especially those who live near or in the mountains. Those who have been close to the area where he walked, they said it was very stinky. My second year hunting with my uncle, while we were walking he stopped all of a sudden and started sniffing the air. During he first snowfall, it would snow and then melt. It would not get very deep at that time. In the olden days, the hunters used to build house wherever they would go hunting. When they knew they were not going to make it back home for a while from the mountains, they used to build their houses with the trees on a slant. They used to stay at a place where they could build a fire under a rock cliff, it was just like a cave, and the rock would come over. One morning when they woke up, they seen footprints going past their camp, they did not hear anything during the night. I have seen it one time when my wife and I were returning from Terrace. We have to tell the stories we know about the hairy man. All of our grandfathers and uncles were hunters and they have told about this. This is all I have to say for now, Mr. Chairman. Horace Stevens, Council of Elders, December 18, 2003.

Emma Nyce: Thank you Mr. Chairman, I am happy to be able to be in attendance and I agree with what Horace has said that we should tell what we know. I have also heard this a long time ago when I was with my grandmother Annie in Greenville when I was small. She told us that our grandfather was not trying to scare us when he was telling us about this in Greenville. I heard William Stevens and Peter Calder telling that they did see it; there were three of them. I forgot the place they had mentioned where they had seen this and we laughed when we heard that they said that he had no clothes on. My grandmother told us not to laugh because it was *hawahlkw (taboo)*. Only those who have cleansed themselves were the ones who seen it; it was not seen by everyone. They seen it at the mouth of the Nass, it was walking along the sand beach. My dad told us that it was across from Mill Bay. He said that it was true and that only those who cleansed themselves were able to see him. He also told us not to laugh at it because it was alive. When I hear children laughing when they hear this story about the Sasquatch, I tell them not to laugh because he will follow them. My dad told us that he *(the hairy man)* was a living creature of some kind. When I heard what Horace had mentioned, I remember what I heard from my grandmother. I know they have seen it at the mouth of the Nass but I have not heard of anyone seeing it up this way. The hunters seen this a long time ago and they say that it is part *naxnok (supernatural)*. My grandfather told this *adaawak (traditional story)*, during the public works. This is all I know about this that is all I will say for now. Emma Nyce, Council of Elders, December 18, 2003.

Peter Clayton: One of my sons used to go fishing trout, they went up. They went three times, one day as they were coming down the trail, they seen a man standing down below. When they had reached down below, the man had disappeared. I told him not to go up there anymore because I remember what my father had told us about the *sbinaxnok (supernatural being)*. I had forgotten some of the story, this is probably the same as what we are discussing now. I will ask my son who was with him at the time so we will know the place where they were. There is another time when I used to gihl'askw *(translation not shown)*. I went with Ester Adams and them and they seen it and she told me not be afraid. This is a very strong *adaawak (traditional story)* that Horace has mentioned. A man can die if he breaks the law regarding the cleansing. I know my son does not have any difficulty whenever he goes hunting. We should ask other elders who have seen a Sasquatch or the *naxnok* and where so that it would become more clear. Somebody saw a man on the other side of Sand Lake. Peter Clayton, Council of Elders, December 18, 2003.

Chapter 2
Bigfoot "Recognition"

Remarkably, in 1975 the United States Army Corps of Engineers officially recognized sasquatch as a "possibility" in the *Washington Environmental Atlas*. According to an article in the *Washington Star News* (July 1975), part of this recognition was based on a Federal Bureau of Investigation (FBI) analysis of a hair sample that could not be found to belong to any known animal. An excerpt from the actual article is shown here.

> **RECOGNITION AT LAST**
>
> Though conceding that his existence is "hotly disputed," the Army Corps of Engineers has officially recognized Sasquatch, the elusive and supposed legendary creature of the Pacific Northwest mountains. Also known as Big Foot, Sasquatch is described in the just-published "Washington Environmental Atlas" as standing as tall as 12 feet and weighing as much as half a ton, covered with long hair except for face and hands, and having "a distinctive human-like form." The atlas, which cost $200,000 to put out, offers a map pin-pointing all known reports of Sasquatch sightings, and notes that a sample of reputed Sasquatch hair was analyzed by the FBI and found to belong to no known animal.

Jean McManus, the editor of the reference work, is quoted as follows in a subsequent newspaper article (*The National Enquirer*, June 11, 1975). "The details about the creature were gathered from many sources – anthropologists, writers and genuine Bigfoot hunters. A great deal of time and effort has gone into hunting this creature, and it seemed only right to include it in the book."

In the same article, Major Fred Shierley, an Army spokesman at the Pentagon, is also quoted: "This is the first time there has ever been official Army reference to the actual existence of Bigfoot. As far as we know, no other government agency has recognized its possible existence before. We believe the Army is the first agency to do it."

The page in the atlas is shown here. The text and chart showing sighting statistics are reprinted on the following pages.

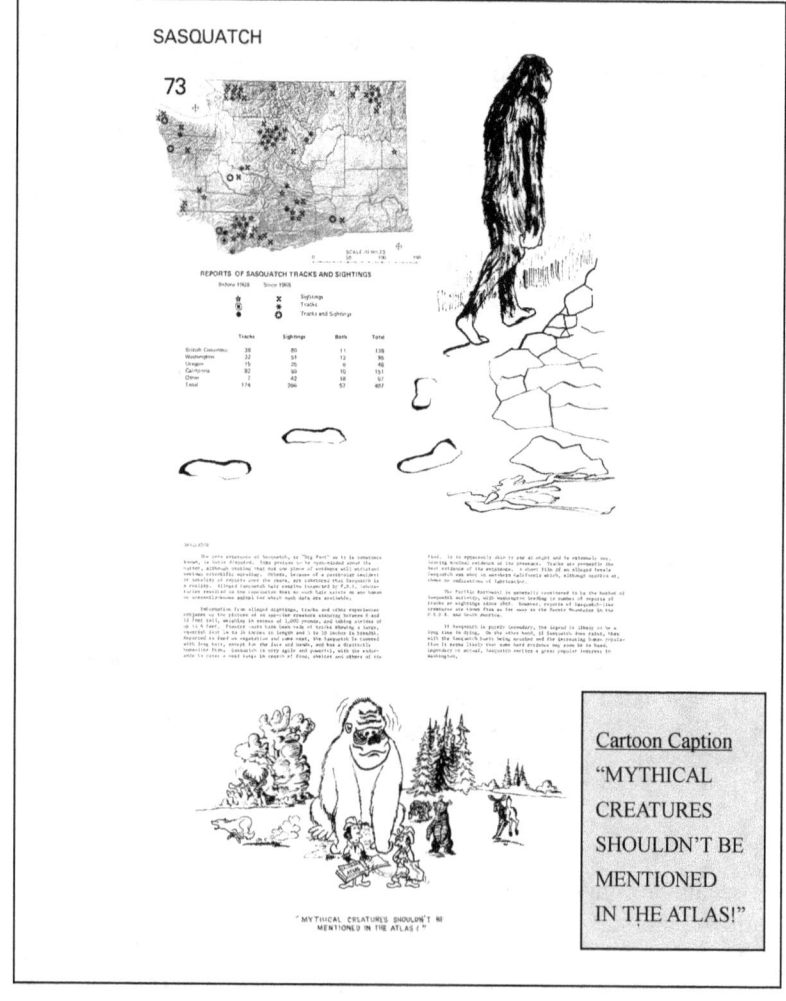

SASQUATCH: The very existence of Sasquatch, or "Big Foot" as it is sometimes known, is hotly disputed. Some profess to be open minded about the matter, although stating that not one piece of evidence will withstand serious scientific scrutiny. Others, because of a particular incident or totality of reports over the years, are convinced that the Sasquatch is a reality. Alleged Sasquatch hair samples inspected by F.B.I. laboratories resulted in the conclusion that no such hair exists on any human or presently-known animal for which such data are available.

Information from alleged sightings, tracks and other experiences conjures up the picture of an ape-like creature standing between 8 and 12 feet tall, weighing in excess of 1,000 pounds, and taking strides of up to 6 feet. Plaster casts have been made of tracks showing a large, squarish foot 14 to 24 inches in length and 5 to 10 inches in breadth. Reported to feed on vegetation and some meat, the Sasquatch is covered with long hair, except for the face and hands, and has a distinctly humanlike form. Sasquatch is very agile and powerful, with the endurance to cover a vast range in search of food, shelter and others of its kind. It is apparently able to see at night and is extremely shy, leaving minimal evidence of its presence. Tracks are presently the best evidence of its existence. A short film of an alleged female Sasquatch was shot in northern California which, although scoffed at, shows no indication of fabrication.

The Pacific Northwest is generally considered to be the hotbed of Sasquatch activity, with Washington leading in number of reports of tracks or sightings since 1968. However, reports of Sasquatch-like creatures are known from as far away as the Pamir Mountains in the U.S.S.R. and South America.

If Sasquatch is purely legendary, the legend is likely to be a long time in dying. On the other hand, if Sasquatch does exist, then with the Sasquatch hunts being mounted and the increasing human population, it seems likely that some hard evidence may soon be in hand. Legendary or actual, Sasquatch excites a great popular interest in Washington.

LOCATION	TRACKS	SIGHTINGS	BOTH	TOTAL
British Columbia	38	89	11	138
Washington	32	51	12	95
Oregon	15	25	6	46
California	82	59	10	151
Other	7	42	18	67
TOTAL	174	266	57	497

The last piece of information in the newspaper article, that concerning analysis of hair, prompted considerable interest among bigfoot enthusiasts. An enquiry by Peter Byrne of The Bigfoot Research Project asking for confirmation and specifics on the analysis received the following response from the FBI.

> Since the publication of the "Washington Environmental Atlas" in 1975, which referred to such examinations, we have received several inquiries similar to yours. However, we have been unable to locate any references to such examinations in our files.

The FBI did, however, follow-up with a Dr. Steve Rice, who was another editor of the Army Atlas. In an official report, the FBI stated: "After checking, Dr. Rice was unable to locate his source of the reported FBI hair examination."

The foregoing information relative to the FBI was obtained by Joedy Cook under the Freedom of Information - Privacy Act. Joedy wrote to the FBI and requested all information relative to bigfoot. He received documentation that primarily deals with the Army Atlas and a subsequent analysis of the hairs submitted by Peter Byrne of The Bigfoot Research Project (examination revealed that the hairs was "of deer family origins"). It is evident the FBI agreed to do this analysis in light of the analysis associated with the atlas. There was nothing else of any significance in the file. However, it appears there is always a little "mystery" whenever the FBI is involved in anything. Attached to the file sent to Joedy Cook was a standard pre-printed "Dear Requester" form. Curiously, a box on this form which states, "See additional information which follows," is checked. At the bottom of the form the following information was manually typed-in:

```
Enclosed are previously processed documents
which relate to "Big Foot." The enclosed are
the best copies available. Serial 4 is miss-
ing from file 95-213013, the file where your
release originates. Our effort to locate
that document was not successful. It is pos-
sible that the number 4 was missed during
the original serialization of the file.
```

    We are left to wonder what was in Serial 4. Certainly the FBI would be a little more efficient in their filing procedures to have omitted this section.

    Nevertheless, we have learned that analysis of hair samples as indicated in the *Washington Star-News* article <u>definitely took place,</u> and that the samples could not be identified. George Clappison did extensive research on this incident and was referred by the FBI to the ex-head of their Hair and Fiber Unit. This person, who now operated his own private laboratory out of his home, was in charge at the time the hair samples were submitted to the FBI. He told Clappison that <u>the analysis was done after hours on employees' own time with the results as indicated.</u> He further stated that no written reports were prepared on the analysis.

    In discussing the whole situation with the current head of the FBI Hair and Fiber Unit, Clappison asked if the unit would now consider analyzing other hair samples. The current manager agreed to perform an analysis; however, he informed the unit would not respond in writing on their findings.

    Unfortunately with hair analysis, we have a "Catch 22" situation. In order to establish that a hair sample came from a sasquatch, it is necessary to compare the sample with an actual sasquatch hair. If the object of the exercise was to prove the creature exists, it would be redundant because its existence would have already been proven by the actual hair sample. The absolute "best" result we can get from analysis of an alleged bigfoot hair sample is to establish that it did not come from any known creature for which a hair sample is on record. The same situation primarily applies to DNA analysis. Here, however, it could be determined if the DNA was from a non-human primate. If a match could not then be found with all known non-human primates, then we would have a good case for bigfoot. To my knowledge, we have not been able to get any DNA from hair samples or feces sample that came from alleged bigfoot.

State - County Initiatives: Bigfoot recognition is somewhat reflected in laws or initiatives aimed at protecting the creature. While Ohio does not have any specific laws, two counties, Skamania and Whatcom, in Washington State have enacted protection legislation. On April 1, 1969 the Board of Commissioners of Skamania County adopted a bigfoot protection ordinance. However, it has been partially repealed and amended because it "may have" exceeded the jurisdictional authority of the Board of Commissioners. The revised ordinance went into effect on April 2, 1984. In 1991, Whatcom County passed a resolution (No. 92-043) which simply declares the

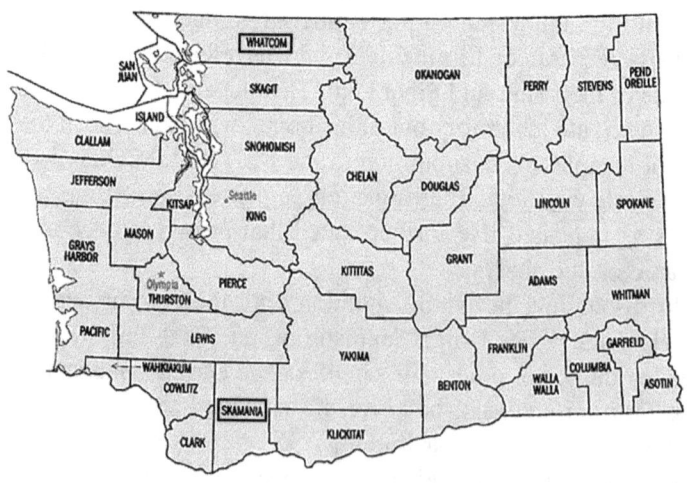

Washington State Counties. Skamania and Whatcom Counties are boxed.

county a sasquatch protection and refuge area.

The full text of the latest (1984) Skamania County ordinance follows.

ORDINANCE NO. 1984-2
PARTIALLY REPEALING AND AMENDING
ORDINANCE NO. 1969-01

WHEREAS, evidence continues to accumulate indicating the possible existence within Skamania County of a nocturnal primate mammal variously described as an ape-like creature or a sub-species of Homo Sapiens; and

WHEREAS, legend, purported recent findings, and spoor support this possibility; and

WHEREAS, this creature is generally and commonly known as "Sasquatch", "Yeti", "bigfoot", or "Giant Hairy Ape", all of which terms may hereinafter be used interchangeably; and

WHEREAS, publicity attendant upon such real or imagined findings and other evidence have resulted in an influx of scientific investigation as well as casual hunters, most of which are armed with lethal weapons; and

WHEREAS, the absence of specific national and state laws restricting the taking of specimens has created a dangerous state of affairs within the county with regard to firearms and other deadly devices used to hunt the Yeti and poses a clear and present danger to the safety and well-being of persons living or traveling within the boundaries of this country as well as the Giant Hairy Apes themselves; and

WHEREAS, previous County Ordinance No. 1969-01 deemed the slaying of such a creature to be a felony (punishable by 5 years in prison) and may have exceeded the jurisdictional authority of that Board of County Commissioners; now, therefore

BE IT HEREBY ORDAINED BY THE BOARD OF COUNTY COMMISSIONERS OF SKAMANIA COUNTY THAT THAT PORTION OF Ordinance No. 1960-01, deeming the slaying of bigfoot to be a felony and punishable by 5 years in prison, is hereby repealed and in its stead the following sections are enacted:

Section1. Sasquatch Refuge. The Sasquatch, Yeti, bigfoot, or Giant Hairy Ape are declared to be endangered species of Skamania County and there is hereby created a <u>Sasquatch Refuge,</u> the boundaries of which shall be co-extensive with the boundaries of Skamania County.

Section 2. Crime – Penalty. From and after the passage of this ordinance the premeditated, willful, or wanton slaying of Sasquatch shall be unlawful and shall be punishable as follows:

(a) If the actor is found to be guilty of such a crime with malice aforethought, such act shall be deemed a Gross Misdemeanor.

(b) If the act is found to be premeditated and willful or wanton but without malice aforethought, such act shall be deemed a Misdemeanor.

(c) A gross misdemeanor slaying of Sasquatch shall be punishable by 1 year in the county jail and a $1,000 fine, or both.

(d) The slaying of Sasquatch which is deemed a misdemeanor shall be punishable by a $500.00 fine and up to 6 months in the county jail, or both.

SECTION 3. Defense. In the prosecution and trial of any accused Sasquatch killer, the fact that the actor is suffering from insane delusions, diminished capacity, or that the act was the product of a diseased mind, shall not be a defense.

SECTION 4. Humanoid/Anthropoid. Should the Skamania County Coroner determine any victim/creature to have been humanoid the Prosecuting Attorney shall pursue the case under existing laws pertaining to homicide. Should the coroner determine the victim to have been an anthropoid (ape-like creature) the Prosecuting Attorney shall proceed under the terms of this ordinance.

BE IT FURTHER ORDAINED that the situation existing constitutes an emergency and as such this ordinance shall become effective immediately upon its passage.

REVIEWED this 2nd day of April, 1984, and set for public hearing on the 16th day of April, 1984, at 10:30 o'clock a.m.
BOARD OF COUNTY COMMISSIONERS
Skamania County, Washington

(Signed by the Chairman, two Commissioners and the County Auditor and Ex-Officio Clerk of the Board.)

Copyright Canada Post Corporation, 1990. Reproduced with permission.

The only marginal consideration given to bigfoot in Canada is the postage stamp (left) issued in 1990. The stamp is in the *Canada's Legendary Creature's* series. In 2005, I had the stamp on the right created under the Canada Post "Picture Postage" process.

Chapter 3

Bigfoot Speculations

While there are many eye-witness descriptions of bigfoot, there are few photographs of the creature. The only reasonably clear images are those of the creature seen in the Patterson/Gimlin film. Working from film frame 352, I performed the following study. The first image (top left) is the actual film frame head which was enhanced with pastels. It will be noted that I closed the creature's mouth to give it a more natural appearance.

Some time later, Yvon Leclerc (Quebec) enhanced the image

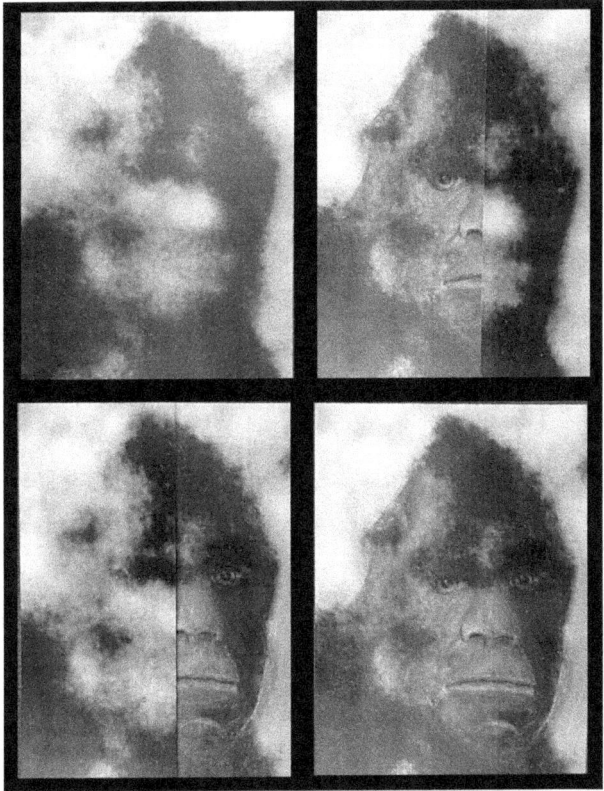

seen in film frame 339. The actual film frame head is shown on the left with Yvon's unique work on the right.

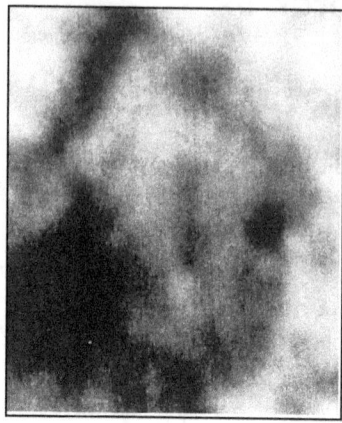 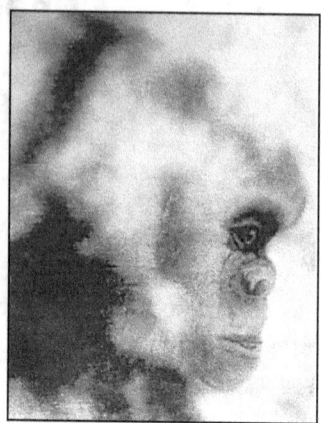

In conjunction with my exhibit at the Vancouver Museum in 2004, I constructed a model sasquatch foot, using an actual 16-inch (40.6-cm) footprint cast to form the sole. The purpose of the exercise was to get a better appreciation of the real size of the actual foot that is making the large footprints.

A sculpture created by Igor Bourtsev in the early 1970s provides

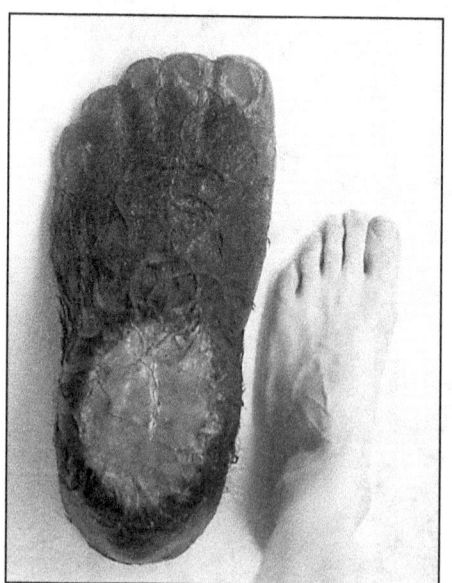 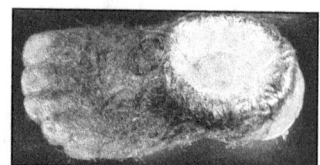 

highly revealing insights on bigfoot's appearance. Igor also based his creation on the Patterson/Gimlin film. I will mention here that what is applicable in the west is just as applicable in the east. North America is just one big territory for bigfoot.

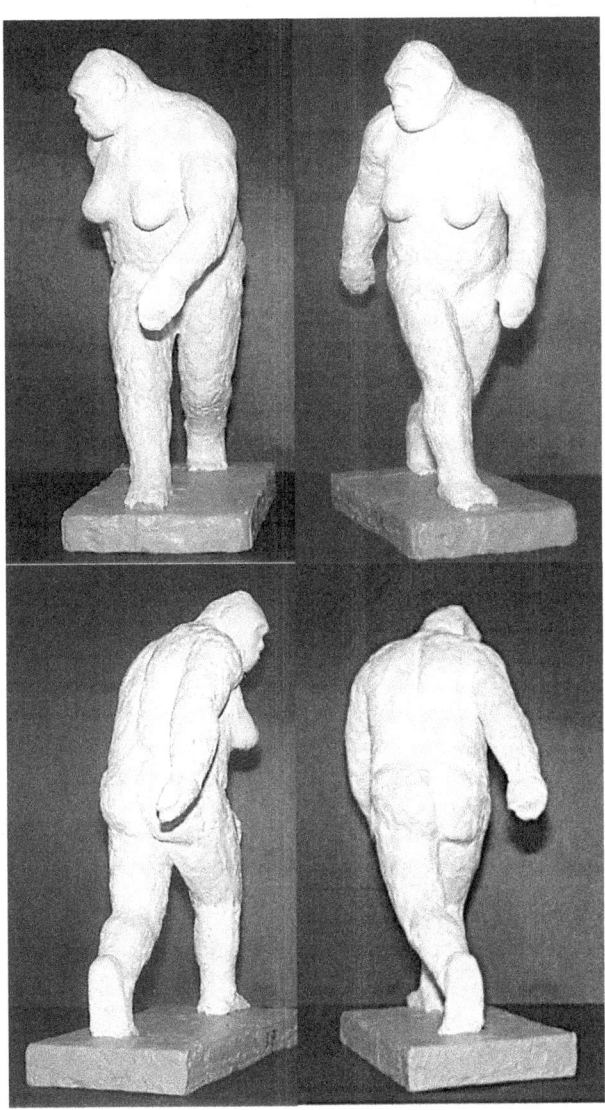

In early 2004, I sought a sculptor to create a bigfoot head for the museum exhibit. A remarkable Vancouver artist, Penny Birnam, was found and she skillfully created four heads, shown below, each with different facial features. It is reasoned that bigfoot, like human beings, would all be different in appearance. The Patterson/Gimlin film provided the inspiration for the sculptures. I provided Penny with the film and a number of enlarged photographs taken from film frames for her study. Penny is a professional artist who specializes in animal sculptures.

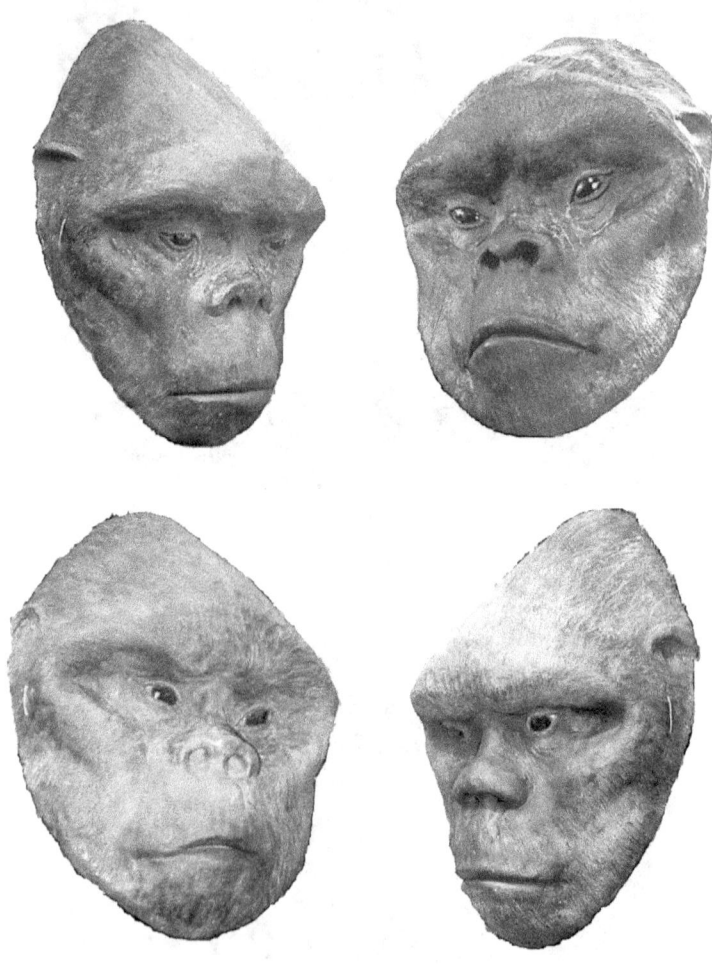

While descriptions of the bigfoot or sasquatch are fairly consistent, some eyewitnesses stress the "ape" side while others stress the "human" side. I created the following sculpture in 2004 giving the latter the most weight. (Size is slightly larger than human.)

This sculpture was created by Roger Patterson in the early 1960s. Its creation signifies Patterson's zeal for bigfoot which totally consumed his life. It appears Patterson envisioned a far more human-looking bigfoot than the one he and Bob Gimlin filmed at Bluff Creek, California in October 1967.

Chapter 4

Ohio Bigfoot in Review

The following are some of the sighting and footprint reports, together with investigation results, on bigfoot in Ohio. The list is by no means comprehensive coverage on this subject. Information was compiled from many sources; however, there are certainly other sources that were not accessed. Nevertheless, it is believed the main or major incidents are included.

It is interesting to note that some early reports often use the term *wild man* to refer to the unusual creature in the account. In some cases, it appears obvious they are referring to an actual wild man or, as it were, a wild human being. In other cases, it does not appear reasonable that the creature was a wild man. Use of the term "wild man" probably came about because early writers really did not know what else to call the creatures. However, when we are confronted with the two terms *wild man* and *gorilla-like creature* in the same account, we can reasonably justify that the creature was what we would now call a bigfoot.

Certainly, only a very low percentage of bigfoot incidents are reported to to research groups for review and possible investigation. Indeed, it appears that many, if not most, incidents remain confidential. They are not even discussed outside a small circle of friends and relatives of the person who had the experience. Sometimes, people come forward with information many months or years after a significant event. Unfortunately, the passing of time clouds recollection. Specific dates, times, locations and sighting details are hazy and cannot be totally depended upon for field investigation.

There is also the problem of resources (manpower and financial) to properly respond to sightings when they are reported. Often, all that can be done is to record the information and keep it for reference. While specific details are missing on some of the following reports, they nevertheless have pieces of information that certainly add to our knowledge of the Ohio bigfoot phenomenon. In many cases, incidents are just referenced. The purpose for including such entries is to show the extent of bigfoot reports in Ohio. The fact that the incident occurred is significant in a work of this nature, even though details are missing. <u>Please bear in mind that your authors do not attest to the credibility of any of the reported incidents.</u> Some accounts certainly appear to be the result of over-active imaginations, and some could certainly be fabrications. <u>In many cases, we fully realize they defy logic</u>.

The following reports are in alphabetical order by county and then chronological order within each county.

ADAMS COUNTY

Rome, 1897: On May 26, 1897 Charles Lukins and Bob Forner claim they encountered a wild man while cutting timber a few miles from Rome. After several struggles, they say they were able to drive the gorilla-like creature into his supposed retreat among the cliffs. They described the "terror" as being about six feet tall (1.83m) and his only covering a mat of long, curly hair. (Source: *The Cleveland Plain Dealer,* May 27, 1897.) *[Comment: The "key" in this historic account that alludes to a bigfoot creature is the compound word, "gorilla-like." Most other accounts of "wild men" do not use this word.]*

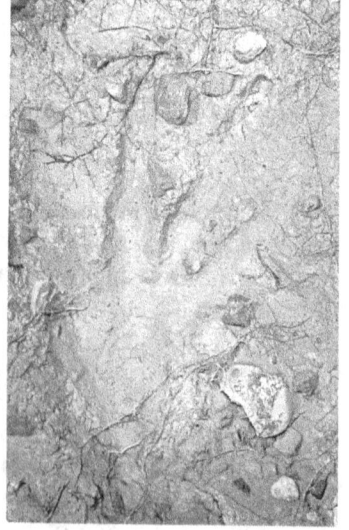

Forest Area, 1995: In May 1995 a hunter was waiting for deer to come by his position when he heard loud thumping. He sensed a strange odor and upon looking in the direction of the thumping, he saw the back of a huge bigfoot-type creature. He noted that the soles of the creature's feet were black.

Investigators went to the area and found what appeared to be a large handprint, as seen in the photograph shown here. The print is about 10 inches (25.4cm) in length from what appears to be the beginning of the palm to the tip of the longest finger. By comparison, a human male hand (man, 6 feet [1.83m], 200 pounds [90.6kg]) would be about 8 inches (20.3cm) in length.

Forest Area, 1996: In July 1996 a retired police officer noticed an unusual odor while out hunting. He looked around and saw a black ape-like creature between 6.5 feet (1.98m) and 7 feet (2.13m) in height standing by a large tree. The creature was looking (staring) at him. He quickly left the scene.

Dunkinsville, 1997: On June 26, 1997 a lady in Dunkinsville states that she saw a strange creature in her front yard. She was watching television at some time between midnight and 1:00 a.m. when she heard her dogs barking outside. She turned on the porch light and observed the creature, which was about 30 feet (9.1m) away. She described the creature as being between 3 and 4 feet (91.4cm and 121.9cm) tall, covered in gray fur or hair about 1.5 inches (3.8cm) long, large, dark eyes, rounded ears that extended above its head, very long arms and a short tail. It made a gurgling sound. The creature looked at the lady for a few moments and then "knuckle-walked" away (like a chimpanzee). It appeared to sort of skip as it moved.

Bentonville, 2002: On February 23, 2002 at about 11:30 p.m., two ladies, Dana Smith and Loren Greer, heard unusual screaming and moaning in the woods near their Bentonville residence. The sounds lasted for between five and ten minutes, and when the ladies looked out into the darkness they saw what they described as a "big man." They telephoned the police who investigated the incident, but did not find anything or anyone. The police subsequently told the ladies to contact Joedy Cook at the Ohio Center for Bigfoot Studies (Cincinnati). The next day, Loren Greer's brother, Tony Greer, scouted the area and found an unusual hand impression and a rough 17-inch (43.2-cm) footprint about 39

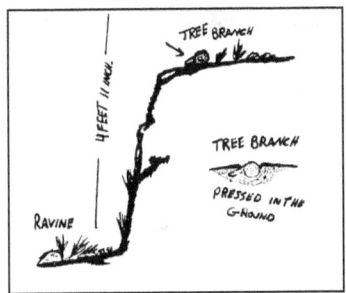

inches (1m) away. Tony, a hunter who has made plaster casts of animal tracks, made a cast of the hand impression. Unfortunately, the footprint was not good enough to cast. He then contacted Joedy Cook who went to investigate the incident. Tony informed Joedy of what he felt were the circumstances and gave him the hand cast. Joedy drew the illustrations seen here. It appears the creature hopped up a little bluff and in doing so, placed its left hand on the ground for leverage. The lower palm of the hand pressed down on a small tree branch that was pressed into the ground. The upper palm and four fingers went into the ground.

As the cast showed dermal ridges (finger prints), Joedy had it professionally photographed (as seen below) and the photographs were subsequently sent to Jimmy Chilcutt, an expert in primate dermal ridges.

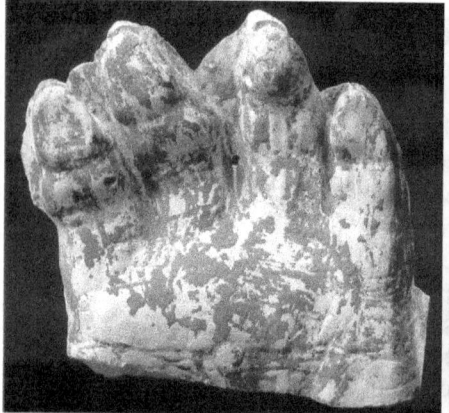
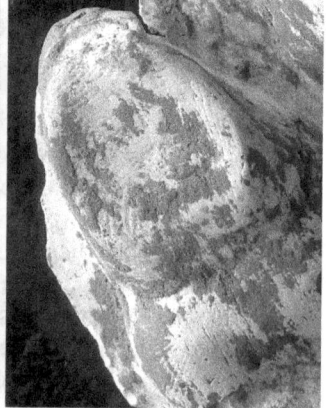

The Bentonville hand cast and enlargement of the index finger (first finger on left). The cast as it is seen here is how it looked when it was given to Joedy Cook. The soil was later cleaned off.

Chilcutt identified the cast as a handprint of an adult gorilla; however, the cast index finger and little finger indicated that the creature was in a postmortem state, or very near death. This conclusion raised suspicion that the print might have been made by a commercially available gorilla hand, such as those available from BoneClones. The BoneClones people were contacted and they provided Chilcutt with cast finger tips from their gorilla hand mould (left hand, same as the cast). The actual cast was then sent to Chilcutt for full investigation and comparison. The following is Chilcutt's official report on the cast and the illustration he provided.

December 16, 2003

After a thorough examination and consultation with Dr. Jeff Meldrum, I have reached the conclusion that the cast is of the left hand of an adult gorilla.

1. The friction ridges visible on the finger tips (A) are in a loop pattern with the core low to the first joint. This is a typical primate characteristic. Also note that the index and little fingers appear to be in a postmortem state.[1]
2. The fingernails (B) are thick, tapered and arced; typical in great apes.
3. The flection creases (C) are similar to humans, but the flection creases (D), called the simian creases, are only found in primates.
4. The texture of the ridges is more consistent with known human and non-human primate friction ridges than that of known Sasquatch friction ridges.
5. The width of the hand is 6 inches (15.2cm) which is larger than any chimp I've printed. But I must confess I haven't printed all the chimps in the world.

These are the factors I've based my conclusion on. It can be argued that since no one has documented a Sasquatch hand print, and this hand print is that of a great ape, then it could be a Sasquatch hand print. My answer to that is that the texture of known Sasquatch foot dermal ridges does not match the dermal ridges on this cast. I have found that the dermal ridge texture in both human and non-human primates stays constant between feet and hands.

The fingers from BoneClones did not match the hand cast.

---

1. This observation is drawn from the fact that two finger tips are concaved (pushed in) which would result if the creature were dead or near death.

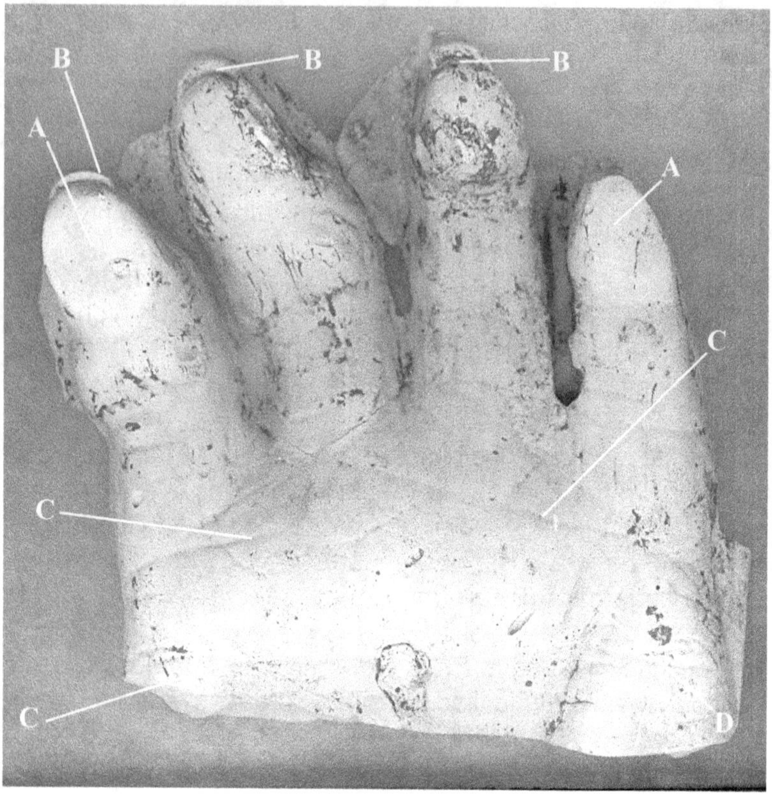

A. Friction ridges in a loop pattern.
B. Fingernails consistent with great apes.
C. Flection creases consistent with great apes.
D. Non-human flection creases (great apes).

Joedy Cook is seen in this photograph holding the hand cast. It has been reasoned that the actual print in the ground (from which the cast was made) had to have been made with a flexible hand. It is clearly seen that the finger tips are concaved (pushed in or flattened out), so this condition rules out a solid hand such as those provided by BoneClones. Subsequent research has led to the speculation that the print could have been made with a stuffed gorilla hand. During the late 1970s and early 1980s, actual gorilla hands were used to make ash trays, and certainly some of these "curios" found their way into the United States. We are convinced that if the hand were faked, Tony Greer was not involved. However, that the print was "planted" and found by chance, is almost as difficult to rationalize as concluding that an actual gorilla (or gorilla-like being) made the print. We can also reason that the story associated with how the print appears to have been made is a little too good for a planned fabrication. The following illustration by Yvon Leclerc further confirms that the cast is that of a gorilla's hand.

Ohio handcast

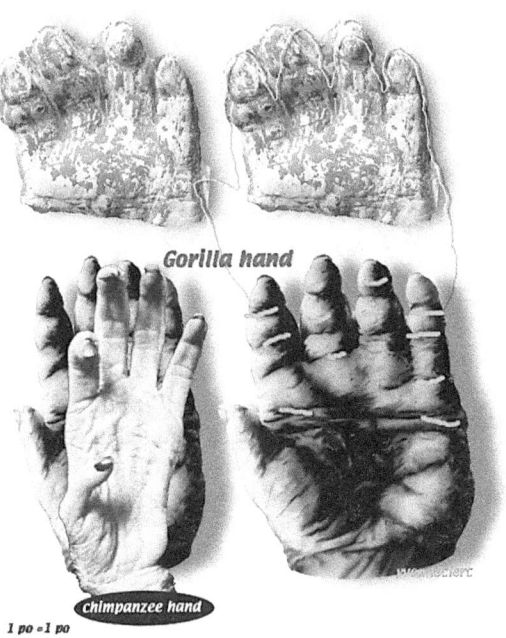

Gorilla hand

chimpanzee hand

1 po = 1 po

The fact that the cast did not show a thumb immediately raised some suspicion that such was done intentionally as part of a fabrication. The position of the thumb on what is believed to be a bigfoot hand cast (adjacent illustration, mirror image) is very different from that of a gorilla. It is reasoned that if a gorilla's hand had been used to make

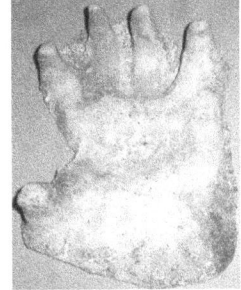

Freeman hand cast, Blue Mountains, Washington, 1995.

the Ohio print (i.e., a definite fabrication) the thumb position, if evident, would have been a "dead give-away." This reasoning, of course, implies that the person who fabricated the print (if in fact it was fabricated) had some knowledge in this regard. If this is the case, I must admit that the reason provided for the missing thumb was very clever. I don't think I would have thought of it.

ALLEN COUNTY

Bluffton, 1956: On November 29, 1956 the following article appeared in the *Bluffton News*. It was reprinted in the same paper in November 1981 with a new title, as follows. The Editor's note reads: *Residents of Bluffton and the surrounding community might remember stories about the "cries like a baby, screams like a woman-mystery animal" from the mid-1950s. Twenty-five years ago this month, in the Nov. 29, 1956,* Bluffton News, *front page, was published this account of the events surrounding the "mystery animal."*

Screaming Mystery Beast Haunted Us 25 Years Ago

A mysterious night-prowling animal that "cries like a baby and screams like a woman" and powerful enough to break a coon dog's back, is again on the prowl in the Lafayette area, according to local coon hunters. The strange animal, that is believed to have killed at least five or six coon dogs since it was first heard in 1950, is back again, according to Mr. and Mrs. Dallas Yoakam of the Lafayette area.

The animal was first heard by the Yoakams while hunting coon in the old Perry Bechtol woods three miles (4.8km) northeast of Lafayette. "First it sounds like the low cry of a human baby," the Yoakams describe it. Their coon hounds approached the sound. Then turned and scurried away, tail between legs. Mr. Yoakam learned that other hunters have heard the same cry. Some said it sounded like an angry woman.

In 1952, one of the Yoakam boys and a friend were hunting when they encountered the weird sound. One of their dogs attacked the animal. The mangled body of the canine was found, neck broken. Attempts to identify and capture the animal have failed.

This year, after Mr. and Mrs. Richard Watt reported hearing the animal near the Sandusky road northeast of Lafayette, a posse was formed and a square mile area was searched without success. The last coon hound reported to have been killed by the animal was owned by Wayne Hauenstein, who lives about seven miles (11.3km) northeast of Lafayette. Mr. Yoakam advises hunters to use caution if they run into the animal. It could be dangerous. Persons having dogs killed under unexplained circumstances are asked to contact Yoakam.

Lima, 1980: In October 1980, a sighting at Lima is referenced (no details). Further, sometime in 1980 in an area west of Lima, unusual animal feces were found by a bigfoot investigator. The feces were sent for analysis which revealed that it contained berry seeds, a large amount of hair and wood fibers. The analysis report concluded that the matter came from a human-type digestive tract. (Source: *Wapakoneta, Ohio News,* March 18, 1981.) *Comment: It appears the analysts felt the feces had to be that of a human being. No explanation was offered as to the unusual nature of the feces' contents for a human.*

Westminster, 1998: In a wooded area near Westminster, a group of four young men experimented with knocking sounds (hitting tree trunks with a branch) at about 11:46 p.m. They received replies to the knocks from different distances and locations. Other forest sounds "spooked" one of the group and he went back to their car. The three remaining individuals later observed a very tall dark figure, at least 7 feet (2.1m) tall watching them at a distance of about 25 feet (7.6m), in front and off to their right. Their attention was redirected from the figure by a loud snapping sound directly in front of them, and when they turned back, the dark figure was gone. *Comment: We are told that one of the individuals had a large steel maglight. Why he did not turn it on and direct it at the dark figure is not known.*

ASHLAND COUNTY

Mohican River, 1997: A student provided the following report of a sighting incident that occurred while he was canoeing with friends along a branch of the Mohican River (near Loudonville) on July 3, 1997.

"I was on a canoe trip with a bunch of friends from school. There were five canoes, each with two people in them. We had canoed about a mile, when we came to a fork in the river. According to the map, the two branches eventually linked up about 1 to 1.5 miles (1.61 to 2.4km) later, so we decided to race. Three canoes went down the fork on the left hand side that had the road next to it. My canoe and another went down the fork that was in the woods. Upon entering the fork, we became aware of a very pungent smell of decaying materials. Soon we were in the middle, and the trees all but blocked out the sky. We stopped for a bathroom break along the river, and on the opposite side noticed a shadow in the trees. We passed it off as a deer. When we continued our trip, we noticed that the smell had gotten closer, and when a female companion glanced towards the smell she silently froze. She said she saw a black 6.5 to 7-foot (1.98m to 2.1-m) creature observing us. We doubled our speed and noticed that it was following us behind the tree line. Soon, we noticed a bend in the river. That is when the creature took off and ran towards the bend. We idled in the river waiting, and then heard a big splash. We began to paddle towards the bend, and noticed just in time to see it coming up onto the opposite bank of the river. It had to cover over 200 yards (182.9m) in 30 seconds. We thought our encounter was over, but later saw it crossing an open clearing casually. That was the last we saw of it. We thought we heard shrieks changing from high to low pitch about 20 minutes after the last visual sighting." *(Aside: The speed indicated equates to about 14 miles [22.5km] per hour.)*

ASHTABULA COUNTY

Andover, 1954: Dean Averick, now a Florida resident, stated that he saw a bigfoot when he operated Dean's Boat Landing at Andover in 1954. The sighting took place at Padanaram, where he stored his boats at the water's edge. Averick related he saw a hairy creature that had a snub nose, peaked eyes and was very chesty. Further, it had light-brown scraggly hair, was over six feet tall (1.83m), and had a long straight back. It walked out of the bush about 125 yards (114.3m) up the shoreline, and waded into about three feet (91.4cm) of water. Averick reasoned that the creature then probably saw him because it immediately headed for a small island off shore. (Source: *Homestead City News,* August 29, 1977.)

Ashtabula City, 1974/79: A sighting between 1974 and 1979 in the city of Ashtabula area is referenced (no details).

Rome, 1980: In June 1980 a family reported seeing an unusual creature at Rome. They first heard "grunting screams," and later found four domestic ducks with their heads bitten off. The creature was described as "gorilla-like." (Source: *The Forest Press*, August 1981.)

Rome, 1981: In June and July 1981, three separate sightings at Rome are referenced. One sighting was in June, the other two in July. The July sightings involved footprints and odors (no details). These latest sightings were investigated by Dennis Pilichis of the Anthropoid Research and Evaluation Center. Pilichis made casts of three-toed tracks found in the vicinity of Johnson Road.

AUGLAIZE COUNTY

Wapakoneta, 1979: In the winter of 1979, Mr. Sheets, a local resident, claims he saw a bigfoot in the Wapakoneta town area at about 1:30 a.m. At this time, Sheets was digging his van out of heavy snow. He observed the creature "with hair blowing" in street lighting by a stop sign. He estimated that its shoulders were about 3 feet (91.4cm) wide and that it stood about 2 to 2.5 feet (61cm to 76.2cm) higher than the stop sign. As the stop sign was 7.5 feet (2.29m) high, this indicated the creature was between 9 and 10 feet (2.74 to 3.1m) tall.

The next day, snowmobilers reported tracks in deep snow about three miles (4.8km) east of the sighting location. The tracks were 18 inches (45.7cm) long and 8 inches (20.3cm) wide, with 7.5-foot (2.89m) strides (i.e., two steps or left heel to next left heel). They were about 8 inches (20.3cm) deep in the snow. The snowmobilers stated that the snow was powdery and couldn't even be packed, which they say indicates that the tracks could not have been faked.

Pictures were taken of the tracks and sent to a computer group in Canada for analysis. The group informed that the tracks were made by a biped, not a man, weighing 800 to 1,000 pounds (362.4 to 453kg) and standing 8 feet 4 inches (2.54m) tall. (Source: *Wapakoneta, Ohio News*, March 18, 1981.)

BELMONT COUNTY

St. Clairsville, 1991: On October 26, 1991, a sighting at St. Clairsville is referenced (no details).

Flushing, 1995: A sighting report of a white bigfoot seen in the Flushing area in early 1995 prompted a research team to visit the area on April 14 of that year. The team found possible bigfoot footprints and a number of unusual conditions at the sighting location. The team re-visited the location on July 6, 1995, and again found possible footprints and additional unusual conditions. The full documentation of the investigations follows.

Flushing Sighting Investigation, April 14, 1995 *(location of sighting of a white bigfoot, investigated by Don Keating)*: Broken trees and broken tree limbs were seen in one area. The limbs appeared to have been broken within a few days. Also, the ground appeared as though it had been trampled. Further, in one spot piles of loose grass and bush appeared to have been placed, and then mashed down forming a crude bed.

As the team progressed further into the bush, they came to a ravine. A set of side-by-side footprints were found that indicated a two-legged creature had jumped to the spot from some distance up the side of the ravine. Where the creature's feet contacted the ground, most of the pine needles were gone, leaving right and left footprints outlined in the needles. The prints were reasonably clear with minor definition indicating toes.

(Note: Don Keating took a video in this area previously and videoed what may be a bigfoot creature, or perhaps two creatures, in a one and one-half second video segment.)

Flushing Sighting Investigation, July 6, 1995 *(location visited in April was re-visited to see if there were any additional anomalies. Also, further investigation was conducted in this whole area)*: On the way to the original site, the group took an old mining road to another location where sightings were said to have occurred. An open area in the woods revealed a number of broken trees and a pile of feathers. The feathers were identified as turkey feathers. They were spread out in a 10 to 15-foot (3.1 to 4.6-m) circle. In a wet, muddy area about 50 yards (45.7m) away, five footprints were found in sequence. The prints measured 15.5 inches (39.4cm) long and about 8 inches (20.3) wide at the ball of the foot.

Although the tracks were vague, toes could be seen on four impressions and right and left feet could be determined. The depth of each track was similar. The fifth track was the least clear, just mashed grass and a broken twig. However, the twig was broken at the position of the big toe, and it was imbedded in the ground with one part pointing upwards.

The tracks were headed in the direction of the area with the broken trees and feathers. Tire tracks were also seen in the mud, indicating recreational vehicles had been in the area. A cougar track was also found in the area, about 4 to 5 inches (10.2 to 12.7cm) wide, indicating a very large cat for Ohio. The team then proceeded to the original site. They did not find any new evidence at this site. One of the team members, while away from the others, reports that he heard rapping noises. Don Keating stated that he had heard the same noise on other occasions, and noted that the number of knocks was always one short of how many people were in a group.

BROWN COUNTY

Rural Area, 1996: In March 1996, a lady on a farm reported that she observed a bigfoot creature in her cornfield. She said the creature was eating corn. When it noticed the lady, it walked away. However, it later returned several times.

Utopia, 1997: On April 6, 1997 a woman in Utopia was awakened about midnight by her three hysterically barking dogs. The dogs then became very quiet, and a minute or so later she heard a terrifying yell, "like a man with a deep grumbly voice." She did not recognize the sound as coming from any animal in her experience – "It did not sound completely human, but it didn't sound completely animal." Right after the sound occurred, the woman noticed a barge on the river (directly outside her residence) that shined a spotlight in the direction of the sound.

BUTLER COUNTY

Hamilton, 1977: In February 1977, footprints measuring 20 inches (50.8cm) long by 6 inches (15.2cm) wide of an unknown creature were found by Jackie Smith. The tracks were in snow beside a mobile home, and they proceeded up a hill with very long strides. They were followed for almost a mile, where they disappeared in bushy woods. Pete Wilson, a newspaper reporter, went to the location and saw the tracks. He stated that he does not think the tracks were faked. (Source: *Jackson Journal Herald,* February 7, 1977.)

CARROLL COUNTY

Minerva, 1978: On September 9, 1978 Henry Colt, who lived about five miles (8.1km) east of Minerva, claims he saw an unknown furry animal in some woods. The creature was squatting next to a tree and it let out a sound similar to a loud cough.

CHAMPAIGN COUNTY

Urbana, 1979: In late July 1979 Ronald Chamberlin claimed he saw an unusual creature at Urbana. Driving through the area about midnight, Chamberlin saw a husky, hairy creature, larger than a dog, standing in the middle of the road (Route 296). Chamberlin stated the creature's hind legs seemed larger than its front legs. It was broad through the shoulders, dark in color, and moved very fast.

Woodstock, 1980: In March 1980 a series of 18- inch (45.7-cm) footprints were found in snow near Woodstock. The adjacent photograph shows one of the prints along with a size 12 shoe print. The prints appear to indicate a very large big toe.

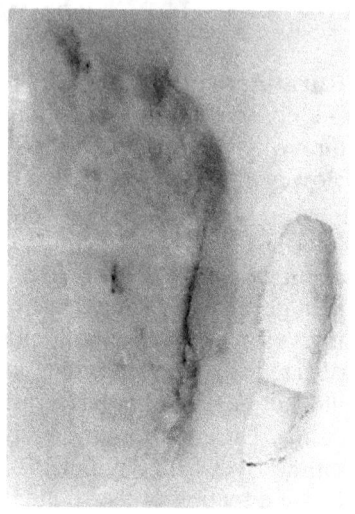

Footprint found in snow near Woodstock, 1980.

Forest Area, 1980s: During the 1980s a group of five people went on a bigfoot excursion in a forested area near Urbana. They heard unusual screams "back and forth" in the woods, and later detected a strange odor. They heard something coming through the woods and splashing through water. They retreated from the area and used binoculars to investigate the scene. They claim that they observed two human-like creatures, perhaps more, which they say were very large – over 8 feet (2.4m) tall.

Rural Area, 1981: On November 11, 1981 a sighting in Champaign County is referenced. Footprints were found in the area (no details).

Woodstock, 1985: In September 1985 a man and his wife claim they saw five bigfoot-like creatures in a field not far from Woodstock. They stated that the group appeared to be a family because the individuals varied in size. They said one of the creatures, presumably a mother, was carrying what appeared to be a baby bigfoot. (Source: *Journal Herald,* Dayton, Ohio, September 14, 1985.)

Woodstock, 1987: In August 1987 Reverend Lee Birt claims he saw a large brown hairy creature in the Woodstock area. Birt found 16-inch (40.6-cm) hair strands that he sent to Ohio State University for analysis. The university reported that it did not know what kind of animal the hair came from. (Source: *Cleveland Plain Dealer,* August 9, 1987.)

CLERMONT COUNTY

Point Isable, 1968: In 1968 a sighting at Point Isable is referenced (no details).

Milford, 1976: On April 4, 1976 a sighting at Milford is referenced (no details).

CLINTON COUNTY

Wilmington, 1961: In 1961 a sighting at Wilmington is referenced. Sounds were reported (no details).

COLUMBIANA COUNTY

Lisbon, 1980: In July 1980 an unusual creature was seen near Lisbon (Roger's Auction area). Residents reported screams and noises. Also, broken tree limbs in the area appeared to have been made by a large creature. Footprints measuring 15 inches (38.1cm) were found and photographed. The photographs were shown to police.

Columbiana City, 1981: In August 1991 a sighting at Columbiana is referenced. Another sighting in the same town took place the following month (no details).

COSHOCTON COUNTY

Rural Area, 1988: On January 1, 1988 a sighting in Coshocton County is referenced (no details).

Forest Area, 1992: A retired machinist who was out hunting encountered a bigfoot in a cave. See Chapter 7, **Ohio Deer Kills and Bigfoot,** for details.

Warsaw, 2000: On May 7, 2000 unusual tracks were found on a farm near Warsaw. On the evening prior to the finding, the resident's dogs started acting strangely. Several times they headed into the woods barking and would then return a short time later. In the morning, the resident went out to continue working on a three-strand wire fence he was constructing. When he arrived at the spot where he had stopped working the previous evening, he found a tree across the lower strands of the fence, forcing them towards the ground. He corrected the obstruction, and then noticed a large footprint on the inside of the fence line. He walked around the area and searched for additional prints. He found another print on a creek sandbar.

Print found near Warsaw in May 2000.

The incident was investigated by Don Keating and the prints were measured and photographed. The prints were about 15 inches (38.1cm) long, 8 inches (20.3cm) at the ball of the foot, and 4 inches (10.2cm) at the heel. The print at the fence, seen here, was less than 100 feet (30.5m) from the resident's house.

CUYAHOGA COUNTY

Cleveland Zoo Woods, 1968: In April 1968 John Keel claims he saw an 8-foot (2.4-m) creature covered with hair in the woods behind the Cleveland Zoo.

Cleveland Zoo Woods, 1972: In August 1972 a 7-foot (2.13-m) creature covered in black hair was also claimed to have been seen in woods behind the Cleveland Zoo. A resident, Wayne E. Lewis, said he encountered the creature at 9:30 p.m., while looking for a kitten. The creature was standing behind a fence and was, according to Lewis, "a lot bigger than he was." Mr. Lewis is 6 feet (1.83m) tall, and said that he weighed 360 pounds (135.9kg). A 19-year-old youth, who also claimed he saw the creature, said it looked like a gorilla, except that it stood straighter. *(Source: Cleveland Plain Dealer,* August 1972.)

DARKE COUNTY

Rural Area, 1981: In August 1981 several residents of Darke County claim they saw a large unusual creature with a strong odor watching them from a cornfield. The group, which was out for a stroll after a family get-together, had previously played a tape of an alleged bigfoot scream. They stood and quietly observed the creature until it moved out of sight.

ERIE COUNTY

Huron Area, 1970: In October 1970 a motorist reported almost hitting a bigfoot-like creature on Fox Road, south of Huron. The motorist described the creature as about 7 feet (2.13m) tall, blackish brown in color, long arms, a pointed head, and deep-set eyes.

FAIRFIELD COUNTY

Stoutsville, 1897: In April 1897 a wild man was reported seen in the woods near the old town of Stout (now Stoutsville). He was described as very tall, half-naked, apparently wearing "tattered pants" and could run like a deer. After a boy was attacked by the oddity, thirty armed men set out to capture it, but it eluded them. (Source: *The Enquirer*, April, 1897.)

Comment: As there is no mention of hair or fur, and "tattered pants" are referenced, this sighting was undoubtedly a case involving an actual wild man. It is curious, however, that with some bigfoot sightings on the Pacific Coast, full details fitting a bigfoot creature are provided, but with the addition that the creature was wearing a loin cloth, or similar covering about its waist.

Pleasantville, 1970s: In the 1970s a sighting at Pleasantville is referenced (no details).

FRANKLIN COUNTY

Dublin, 1972: A sighting in November 1972 is referenced. Footprints were found in the area (no details).

Jack Nicholson Golf Course, 1973: In October 1973 a number of witnesses, including two security guards, reported seeing an 8-foot (2.4-m) hairy monster near, and later actually on, the Jack Nicholson Golf Course, which is located northwest of Dublin. A spokesman for the Franklin County Sheriff's Department said that the monster had been spotted three times by the guards – once standing in the roadway, next in a cemetery near the golf course, and the last time on the golf course by a tree. It fled silently when it saw the guards. A sheriff's lieutenant and sergeant spent several hours investigating the incident and questioning the security guards, who were reportedly very frightened.

The guards were employees of Able Detective & Security Systems. They were assigned to guard the golf course, which was under construction at that time. The guards were issued rifles after a farmer earlier reported seeing the monster near a course fairway. A footprint found along the creek bank resembled a human foot, except that it appeared to have three toes or claws. The entire footprint was about 12 inches (30.5cm) long and 7 inches (17.8cm) wide. The security firm supervisor said that after the monster was first sighted, he doubled-up the guards on duty at the course. (Source: *Michigan Anomaly Research,* October 15, 1973.)

Westerville, 1974: In September 1974 footprints 14 inches (35.6cm) long and 6 inches (15.2cm) wide, showing five toes, were seen and photographed in a field close to woods near Westerville.

Rural Area, 1997: On a Saturday morning in June 1997 a woman driving on Willams Road (near Alum Creek Drive) saw an unusual creature crouched down on the road side. As the woman approached the creature it stood up on two legs. It looked at the woman momentarily, and then stepped over the guard rail and disappeared in a wooded area. The woman described the creature as reddish brown in color with a very large hairy head – twice as big as a bear's head. The woman did not inform the news media of her experience. One week later, Channel 10 in Columbus, Ohio reported a sighting by someone else in the same area.

GALLIA COUNTY

Gallipolis, 1869: On January 23, 1869 the *Minnesota Weekly Record* featured the following article:

A Gorilla in Ohio

Gallipolis is excited over a wild man, who is reported to haunt the woods near that city. He goes naked, is covered with hair, is gigantic in height, and "his eyes start from their sockets." A carriage, containing a man and daughter was attacked by him a few days ago. He is said to have bounded at the father, catching him in a grip like that of a vice, hurling him to the earth, falling on him and endeavoring to bite and scratch like a wild animal. The struggle was long and fearful, rolling and wallowing in the deep mud, half suffocated *(the father)*, sometimes beneath his adversary, whose burning and maniac eyes glared into his own with murderous and savage intensity. Just as he *(the father)* was about to become exhausted from his exertions, his daughter, taking courage at the imminent danger of her parent, snatched up a rock and hurling *(it)* at the head of her father's would-be murderer, was fortunate enough to put an end to the struggle by striking him *(the wild man)*, after which he slowly got up and retired into the neighboring copse that skirted the road.

Comment: This report is of a violent nature which is quite unusual. Most sightings involve a somewhat benevolent creature that quickly but calmly walks away when confronted with human beings. The most noted case of violence is that recounted by Theodore Roosevelt in his book Wilderness Hunter, Outdoor Pastimes of an American Hunter, *(G.P. Putnam's Sons, 1892). The story, which was related to Roosevelt while he was hunting in the Bitterroot Mountains (Idaho-Montana border), tells of a possible bigfoot that killed a man.*

Rural Area, 1912: In 1912 an unusual incident occurred on a Gallia County farm. A man from this area, who was 5 or 6 years old at that time, recalled the following: "I was out picking berries with my mother on the farm. When we started home, we went straight north to the woods and there was a path that went east and west. Well, we went east and when we got to the woods, my mother said 'look there' and we heard something and saw something and it was a monster. She asked, 'Did you see that? Let's go!' We started east on the path, which was at top of a ridge. We started off at a pretty good clip, and she somehow got 30 to 50 feet (9.1 to 25.2m) ahead of me. She started screaming at me. Well, the monster was standing just a bit to our right when we first saw it, and it started walking parallel to us. It was over a roll in the land that I couldn't see it from its knees down. I don't remember about the arms. It had a bulky head, appeared to have no neck to it, and monstrous wide shoulders; probably twice the size of a man's shoulders. It had walked a little bit to the left of me and it was growling or barking. We got to the top of the ridge, about 100 yards (91.4m) and the thing was gone."

Comment: The use of the word "monster" in this account, together with the estimated size of the creature, would probably identify it as a bigfoot.

Gallipolis, 1969: On January 23, 1969 a sighting at Gallipolis is referenced (no details).

GREENE COUNTY

Xenia, 1986 and 1989: In 1986 and 1989 sightings at Xenia are referenced (no details).

GUERNSEY COUNTY

Rural Area, 1992: In December 1992 a Mennonite farmer reported seeing and hearing a bigfoot near his farm over the course of a few weeks. See Chapter 7, **Ohio Deer Kills and Bigfoot,** for details.

HAMILTON COUNTY

Cincinnati, 1959: In early 1959 a trucker telephoned police and reported that he saw a hulking creature climb out of the Ohio River and onto the shore at Cincinnati. It was a cold night and a violent wind was whipping the river into 6-foot (1.8-m) waves. Later, another call was received by the police. Across the river at Covington, Kentucky, a motorist reported seeing a thing on two legs, three or four times the size of a man and much bulkier, on a bridge over the Licking River. (Source: *Cincinnati Post and Times Star,* February 3, 1959.)

Colerain Township, 1981: In 1981 a young married couple, Jim and Kathy, were horseback riding in Colerain Township (gravel pit, near East Miami River Road). They proceeded down the pipeline access, which is wooded, hilly, and rough. When they had descended the last big hill, they came to a dirt road with lakes on both sides and a gravel-pit nearby. Here, they unsaddled their horses and rested for a while.

As the mosquitoes were so thick, they soon re-saddled and continued down the dirt road towards the gravel pit. Jim was riding slightly ahead of Kathy on the larger of the two horses. The couple heard an unusual heaving breathing sound, but did not pay any attention to it. They then claimed that they saw a dark upright-walking figure coming towards them. Using the size of the larger horse as a standard, they estimate the creature was about 8 feet (2.4m) tall. The creature was making the unusual breathing sound that the couple had just previously heard. Both horses were immediately spooked, and became virtually uncontrollable. The horses turned and ran over the bank and bolted for about 1.25 miles (2km). They calmed down somewhat at this point, but continued to fight their riders for the next 2.5 miles (4km) or so. The couple subsequently reported the incident to the police, who did not take the couple seriously. Nevertheless, the police did a cursory check of the area. Kathy pointed out that the horses were well-trained show animals. From her experience with horses, she emphatically stated that they do not spook when they meet normal creatures. Kathy stated that what she and her husband saw was definitely not normal.

Cincy-Lunkin Airport, 1984: In 1984 a sighting near the Cincy-Lunken Airport is referenced. Footprints were found in the area (no details).

Sharonville, 1984: On the evening of July 2, 1984, a mother and her three children saw a bigfoot in Sharonville. The mother states the creature had a grin on its face as it watched her children. The drawing shown here depicts what the witnesses say they saw.

Sedamsville, 1995: In September 1995 a resident of Sedamsville was walking beside the Ohio River on his way to do some night fishing. As he walked down a bushy trail, he claims that he saw a large hairy man-like creature directly in front of him. The resident had a glass bottle in his hand and he threw it at the creature before running away. The resident went into a truck stop and told his story to several truckers, who went back to the sighting location with him. The area was searched but nothing was found.

Madisonville, 1995: In the winter of 1995 a series of tracks about 9 inches (22.8cm) long and 5.5 inches (14cm) wide were found in a wooded area near Madisonville. As these tracks were in snow, some of them were quite clear, showing very definite toe impressions. The following photographs show the tracks and a single print.

Comment: An average 9-inch (22.9-cm) human foot would be no more than 4 inches (10.2cm) wide. Certainly, whatever made the tracks walked on two legs. Also, the prints indicate a very straight walking pattern, which is characteristic of bigfoot tracks found in the West. While a 9-inch (22.9-cm) print would not indicate a very tall creature, smaller tracks have been found among very large tracks in other finds. This implies there are smaller (perhaps younger) creatures that travel with larger creatures.

Forest Area, 1995: Researchers found unusual dome-shaped structures made of forest material. A similar structure was found earlier this year in Summit County, where alleged bigfoot activity had possibly taken place. Refer to the entries for Akron under Summit County for details on all structures found.

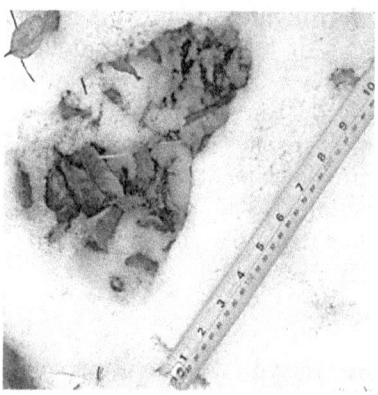

Series of tracks (top), and a single track from that series (lower), found near Madisonville in 1995.

HOCKING COUNTY

Logan, 1897: In May 1897 a newspaper reported that farmers near Logan were greatly excited over the appearance of a strange animal in that vicinity. Numerous sheep and lambs had disappeared.

Several old pioneers, who had heard the cries of the beast at night, said it was a panther, while others said the cries resembled that of a wild cat. Since numerous sheep were missing, the farmers decided to go on a grand hunt for the beast. Every man who was able to carry a gun was to be pressed into service, "with view of effecting the capture of the strange visitor." *(Source: The Akron, Ohio Beacon & Republican – Buckeye News, May 26, 1897.)*

Comment: Here, we have virtually nothing to go on to determine the nature of the "strange animal." All we know is that the creature's cries were different from other animals in the area, as they could not be positively identified.

JEFFERSON COUNTY

Bloomingdale, 1979: In January 1979 Beverly Fletcher, a microbiology student at Ohio State University, found three large human-like footprints in light snow near the wooded area behind her parents' home in Bloomingdale. Two of the prints were side-by-side, and the third was about 3 feet (91.4cm) ahead of the other two. They led into the wooded area. The prints, which were 15 to 18 inches (38.1 to 45.7cm) long, appeared to have clawed toes. The claws had dug into the ground. The snow in the area of the prints was not disturbed except for the paw prints of a small dog.

Miss Fletcher took a Polaroid picture of the clearest print as seen here. She and her mother state they have never seen similar tracks, nor have they ever seen bear or even deer tracks around their home. The area of the prints was examined the following week after the snow had melted. Indentations in the ground that corresponded to the claws in the prints were visible. (Source: *News Register,* Bloomingdale, January 1979.)

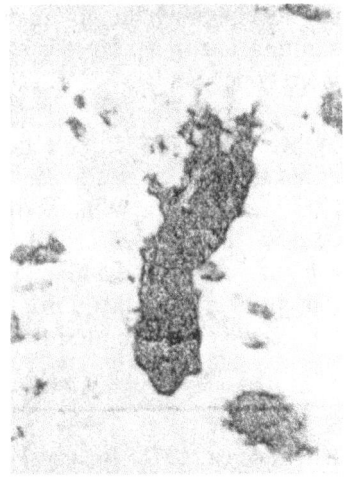

One of three prints, with what appeared to be claws, found in Bloomingdale in 1979.

Comment: In 1979 over fifty percent of Jefferson County was forested, which provides some credibility for the existence or passage of bigfoot in this county. However, the fact that the Bloomingdale prints show claws is unusual. Normally, the toes on possible bigfoot prints are very similar to human toes with no indication of claws. Nevertheless, nails on a bigfoot could eventually become claws if allowed to grow unchecked, that is, worn, bitten, or broken-off.

Smithfield, 1988: During the winter of 1988 (or thereabouts) a father and his son observed two hairy white creatures near Smithfield. The incident occurred along the segment of road that connects the gates of Friendship Lake to Smithfield. The creatures were seen in a field on the right side of the road. They were larger than a man, both in build and height. They were walking down a hill towards a pond and were carrying something that appeared to be about the size of the bed of a pick-up truck. When the creatures saw that they were being observed, they dropped what they were carrying and fell to the ground. They then tried to hide in the bushes, but could still be seen. The father and son observed the creatures for one to two minutes. They then decided to turn their truck around for a better view, so drove about 220 yards (201m) and doubled back, taking not much longer than a minute. Upon arrival back at the sighting location both the creatures and what they were carrying were gone.

Wintersville, 1997: In the fall of 1997 a twelve-year-old boy sighted an unusual creature while squirrel hunting with his grandfather near their home at Wintersville. The boy left his grandfather sitting on a log, and moved off into the forest. He saw a dark hairy figure run out from behind a tree. He states it was bigger than a person and ran upright, "very fast." Both the boy and his grandfather had heard knocking sounds coming from the woods on previous occasions.

KNOX COUNTY

Rural Area, 1978: In July 1978 a Knox County deputy sheriff proceeded to investigate a traffic accident on Ohio Highway 95 near the Richland County line. Upon arriving on the scene, the deputy found a badly frightened young man who was so scared that he scared the deputy. The man told the deputy he ran off the pavement to avoid hitting a bigfoot The man said that the creature was as big as a bear on its hind legs. He urged the deputy to check the adjacent woods. The skeptical deputy simply began his routine accident investigation, despite the protests of the man who felt that the the deputy should have proceeded immediately into the woods.

The newspaper that carried this report stated that since this incident, three other sightings had occurred in Richland County of a creature 7 to 9 feet (2.13-2.74m) tall, with red eyes and a head as large as a tractor tire. The last sighting was by a young girl who was badly shaken afterwards. The Knox County deputy who handled the Ohio Highway 95 accident stated he isn't so sure that the man in the accident was lying, because he saw the creature before the sightings in Richland County.

It was also mentioned that there was an unconfirmed report of a bear captured in Ashland County. If a bear was in the southernmost area of that county, it was reasoned that it could have been responsible for the sightings. (Source: *Mount Vernon News,* July 18, 1978.)

LAWRENCE COUNTY

Hanging Rock, 1940s: During the mid 1940s, a family (father, mother and daughter) were driving through Hanging Rock on Highway 52. They saw a reddish-brown ape-like creature standing on Hanging Rock Hill looking down at the highway.

Rural Area, 1972: In November 1972 a cab driver reported to police at Ironton that he saw a large white ape-like thing crossing a highway dragging a dog or a deer. Ironton is on the Ohio River at the south edge of Wayne National Forest. (Source: *Cleveland Plain Dealer,* November 30, 1972.)

Lake Vesuvius, 2001: In July 2001 James Good was fishing at Lake Vesuvius. He states that he saw an unusual creature with long arms and no neck walking across a nearby field. He estimated its height to be at least 8 feet (2.4m) and its weight at least 600 pounds (271.8kg). He was close enough to the creature to detect an odor, which he described as "like a wet dog," and to notice that the creature's nose was, in Good's words, "human." He also commented on the creature's unusual walk. The creature just kept moving and soon disappeared into a wooded area. Good

walked over to the wooded area to see if he could see the creature again, but it was nowhere in sight. However, he noticed that that it left footprints in the soft wet ground, so he went to get his camera. Shown here is a photograph he took of one of the prints.

Comment: Remarkably, it appears that the prints are only about 10 inches (25.4cm) long. A creature of the height and weight mentioned would certainly have a much larger foot. Nevertheless, the prints are not very distinct at the toes. As can be seen the ground was very wet and as such would quickly distort the prints. It might be noted that Lake Vesuvious is in a large wilderness recreation area by the same name. A short distance east is Wayne National Forest.

LOGAN COUNTY

North Lewisburg, 1979: On October 7, 1979, Scott Simpson, age 15, was out walking near his home in North Lewisburg. He headed down a lane between two fields that led to a wooded area. He claims that he saw an unusual creature about 500 (152m) feet ahead of him standing on a slight rise at the end of the lane. The creature screamed and ran into the woods. Simpson described the creature as 5 to 6 feet (1.5 to 1.8m) tall, standing on two legs, slumped at the shoulders and black in color. He informed his parents of the sighting and they went with him to the woods to investigate. They found fresh footprints and older prints, all measuring 10 inches (25.4cm) long and 4 inches (10.16cm) wide. The prints were pigeon-toed and flat-footed. A neighbor reported hearing an unusual scream at about the same time Simpson saw the creature.

The sighting was later (date uncertain) investigated by Rev. Lee Birt, accompanied by Larry and Peggy Tilman. A small clump of short hairs .375 inches (9mm) long, terminating in flesh or skin, and long hair strands were found on a fence that had been knocked down.

North Lewisburg, 1980: Sightings at the Simpson farm near North Lewisburg sometime in 1980 and July 1980, are referenced (no details).

West Mansfield, 1980: In June 1980 a farmer reported seeing a 7-foot (2.1-m) hairy animal near West Mansfield. The farmer was unloading pigs at the time and saw the creature near his barn. It quickly ran away when noticed. Another sighting that same month and year near West Mansfield is referenced (no details).

Russell's Point Area, 1980: In June 1980 a six-foot-tall (1.85m) creature that had a horrible smell is said to have frightened a farmer and his dog in the Russell's Point area. The creature left 16-inch (40.64cm) footprints.

Rural Area, 1980: About June 22, 1980, some residents claim they saw a hulking shape rush into the woods late one night. (Source: *The New York Globe* June, 1980.)

On June 25, 1980, just after 8:00 a.m., resident Tom Quay went out to his mail box. As he returned to his house, he claims he saw something at the edge of the woods about 120 yards (109.7m) away. Quay described what he saw as being 9 to 10 feet (2.74 to 3.1m) tall, black in color, hairy and having little or no neck. He hurried into the house and called the sheriff's department. Sheriff's deputies soon arrived on the scene. Trampled corn and crushed grass indicated the recent presence of some kind of creature. (Source: *The Chronicle Telegram*, July 2, 1980.)

On June 26, 1980, resident Larry Ramey claims he saw an unusual creature while driving a tractor on his farm. Ramey states that the creature came out of the woods, saw the tractor and came towards it. The creature went back into the woods when another tractor approached.

North Lewisburg, 1981: In March 1981 another sighting at North Lewisburg is referenced (no details).

Bellefontaine, 1981: In 1981 some people in Bellefontaine were having a housewarming party and were taking snapshots of the event. When the prints were developed, one picture showed a bigfoot creature looking in the window.

A month after the picture was taken, one of the men (a deputy sheriff) who was at the house party walked out behind his barn and claims he saw first one bigfoot creature, and later a second different creature. The deputy was wearing his pistol when he saw the first creature, but he didn't want to shoot it. He went back into the house, got his shotgun and a flashlight, and then went around the other side of the barn, hoping to be further away from the creature (this was done so that the deputy would not be noticed). When he was in position, he saw the second creature. We are left to wonder as to what the deputy then observed. As for the photograph taken at the first incident, we are informed that it was subsequently destroyed. Nothing is mentioned of the negative, if such existed.

The Ohio Mammal Research Team went to investigate the scene when informed by the deputy. The team members claim they saw three bigfoot that night. The next day, they made plaster casts of footprints found in the area. According to the team, the evidence collected points to a family of four bigfoot inhabiting that particular area of Bellefontaine.

In another incident *(date and location not given)*, the team found fresh evidence that some kind of creature had moved just ahead of them. They moved to the edge of a hill to get a better look, and saw something moving in the bush about 100 yards (91.4m) away. All of a sudden it moved out in the open, and then off into the thicket. The team stated that all they could do was watch the creature as it moved off at a speed which they could not match. John Thomas, a member of the Mammal Research Team, estimates there are about 1,000 bigfoot living in Ohio. (Source: *Wapakoneta, Ohio News,* March 18, 1981.)

Comment: Unfortunately, the team did not, or could not, get photographs of their Bellefontaine sightings. We have no explanation from the group in this regard.

North Lewisburg, 1995: The sightings at the Simpson farm near North Lewisburg during the 1980s prompted researchers to visit the farm. The full documentation of the investigation follows:

North Lewisburg Sighting Investigation, March 23, 1995 *(Simpson Farm, scene of a number of sightings during the 1980s).* Unusual animal feces were found on a tree limb. The limb was about 6 inches (15.2.cm) in diameter, branching out from the main trunk at about 3 feet (91.4cm) off the ground. The limb had been broken as if pulled down or weighted down.

The amount and size of the feces was unusual. The size was about 1.5 inches (3.8cm) in diameter. The matter appeared to contain a large amount of partially digested corn together with hair fibers. Mrs. Simpson stated that corn was not grown in the immediate area. The height at which the feces were found ruled out known animals that were in the area.

All around the tree a substance resembling chewed berries was found. The substance appeared to contain seeds like a tomato. It was noted that there were no berries in the area at that time of the year.

A large amount of white hair with dark tips, 2 or 3 inches long (5.1 or 7.6cm) was found in the immediate area. It is possible this hair belonged to an animal that had been eaten. No other animal parts were found. A footprint about 10 inches (25.4cm) long with large toes was found nearby. Some of the hair mentioned was imbedded in the print (see first photograph).

Additional footprints were found in the immediate area. The toes and ball of the foot were very evident in these prints, most of which measured 10 inches (25.4 cm) long (see second photograph). One print measured 18 inches (45.7cm) long. There were some three-toe prints and some five-toe prints. It was estimated that the prints were less than six months old.

A lot of broken trees were also observed. The trees had been snapped at a height of between 8 and 12 feet (2.5 and 3.7m). A number of smaller trees had been bent down forming a hollow.

Footprints found by investigators at the Simpson farm.

LORAIN COUNTY

Oberlin, 1973: In August 1973 at about 2:00 a.m., Rudy Randolph and five other raccoon hunters claim they saw an 8-foot (2.44m), shaggy-haired, stinking animal with red glowing eyes at Oberlin. The hunters' dogs crowded the creature into a cornfield where it apparently vanished from view. However, as the men returned to their cars, the creature chased them. Police investigated and found footprints in the vicinity, but they were not clear enough to cast. Nevertheless, it was determined that the creature was bipedal and had a 5-foot (1.5m) step. (Source: *Cleveland Magazine,* October 1973.)

MADISON COUNTY

London, 1979: In 1979 six people, including a local church minister, went out to investigate strange sounds in woods near London. They claim to have observed what they think was a young bigfoot. They described the creature as being about 5 feet (1.52m) tall and covered in white hair. When they viewed the creature with binoculars, they observed nearby what appeared to be a larger bigfoot creature, black in color, crouched down in the long grass. The group had a camera and took pictures of what appeared to be footprints. They did not report getting a photograph of the creature (or creatures).

London, 1980s: During the 1980s, the following incidents in the London area were reported:

While working late one night on his car, a man heard an unusual sound coming from outside his garage. He described the sound as similar to paper being torn. Thinking it was probably his brother-in-law playing a joke, the man carried on with his work. When he looked up, he claims that he saw a human-like creature covered in dark hair standing in the garage doorway. The man ran into his house and got his gun. When he returned to the garage, the creature had vanished. Considering the height of the garage doorway, the man stated that the creature was very tall.

Also in 1980 large human-like footprints were reported found alongside woods in the London area. Investigators went to the scene and photographed 19-inch (48.3-cm) prints. Other prints were found in the woods.

Another 1980s incident involves a mother who states that a bigfoot would lie on the bank at the end of her yard and watch children at play.

West Jefferson, 1980s: During the 1980s, a lady in West Jefferson claims she and her family saw a bigfoot on a number of occasions in woods near her residence. She placed food atop a stepladder in her back yard to attract the creature. She positioned the ladder in a garden in order to get footprints. The creature took the food and left prints, which she photographed. (*Comment: The photographs have not come to light, to our knowledge.*)

Plain City, 1980: On July 15, 1980 Charles Lovejoy, Ron Winn, and friends saw what they believe was a bigfoot at about 8:30 p.m. off Smith-Calhoun Road near Big Darby Creek. The group was parked near the creek, and when they turned up the volume on their stereo, a huge creature stood up in a clearing about 35 yards (32m) away. The creature was about 7 feet (2.1m) tall with wiry black hair covering its body. Its face was covered with a lighter colored hair. It had humanoid features. The creature disappeared and the group went over to where it stood to see if they could find any footprints. They stated they again spotted the creature and then left the area. *(Columbus Dispatch, July 20, 1980)*

London, 1981: In 1981 four people in the London area drove into local woods to investigate strange lights over the area. They had found unusual footprints and heard strange noises in these woods on previous occasions. They parked their car and looked around on foot, whereupon they heard something coming through the woods. They became frightened and quickly returned to their car.

As the driver started the car, the group claims that the back of the vehicle was lifted off the ground. The driver tried to accelerate, but as the back wheels were elevated, the car would not move. The driver observed in his rear-view mirror that some sort of creature was lifting his car. When the creature finally let go of the car, it tore off the license plate.

The next day, the driver received a call from police informing him that his license plate had been found in the country. The driver states that the plate was crumpled up like a piece of paper.

INDIANA CAR INCIDENT

There have been other reports of bigfoot creatures damaging cars. A highly noteworthy incident occurred in Ohio County, Indiana in April, 1977. In this case, it is claimed that the creature crashed against the side of a parked vehicle causing a dent. Two adults (husband and wife) were in the car at the time, just in the process of getting out of the vehicle at their home. The creature was described as black, hairy, ape-like, red eyes, human-shaped head, very long arms and about 12 feet (3.66m) tall. It emitted a strange cry as it approached the car. As the husband quickly drove off, the creature chased the car, swinging its arms. The police were notified and went to the scene, however, no evidence of the creature was found. The same couple saw the creature again the following night, when again getting out of their car at their home. This time, the husband was armed with a .22 caliber rifle. The couple first heard the strange cry they heard the previous night and then observed the creature perched on a hill. The husband fired 15 rounds at it. The creature dove to the ground and disappeared. The police were again notified and went to the scene to investigate. Again, no evidence was found.

In July 1981 a sighting at London is referenced (no details).
West Jefferson, 1981: On July 16, 1981, a sighting at West Jefferson is referenced (no details).

London, 1985: In 1985 four boys in the London area (ages 10-14) skipped church on Sunday and went down by the railroad tracks. They went into a berry patch and observed at distance a very tall man in another part of the patch. The man appeared to have a reddish beard.

The boys hid, and the man came close to their position. When he was in full view, they saw that he had hair all over his body. The boys estimate that he was about eight feet (2.44m) tall. The creature noticed the boy, looked at them for a moment and then ran up the railroad tracks.

The boys again observed the creature at a later date and started to follow it. However, they became frightened and turned back.

On yet another occasion, the boys state they saw the creature while they were playing ball. One of the boys hollered at it and the creature growled at them and left. Investigators went to the scene and found unusually large footprints.

West Jefferson, 1985: In September 1985 Betty Powell, a resident of West Jefferson, states that huge ape-like creatures frequently took up residence in a wooded area near her home. Powell claims she fed three of the creatures dog food and deer meat. She also claims to have seen a UFO land near a local creek. *(Note: There is no indication that the two events are connected.) (*Source: *Jackson Journal Herald,* Dayton, Ohio, September 14, 1985.)

Comment: In all we have three ladies who claimed they have provided food for bigfoot creatures: the lady with the ladder, Betty Powell, and Mrs. Cayton in Stark County (see entry for this county below). We find these claims difficult to believe. It is certainly possible these ladies put out food, and it is equally possible that the food disappeared. Any creature, however, could have taken the food. It cannot be substantiated that it was taken by a bigfoot without photographic evidence.

MUSKINGUM COUNTY

Hopewell Township, 1897: In June 1897 Hopewell township residents reported the appearance of a nude man in Ogle's Woods, from which he approached the public road twice in two days. He ran after an employee of Henry Creeger, who was driving *(early automobile assumed)*, but fearing the nude man was demented, the employee drove rapidly and was soon out of sight. The wife of Rev. G.A. Bartlebaugh, accompanied by her little son, was returning to the city from the country when the man appeared wearing nothing but a hat. Mrs. Bartlebaugh became frightened and turned around and returned to the city over another road. Search parties attempted to capture the man and discover whether he was wild or insane. (Source: *The Cleveland Plain Dealer,* June 16, 1897.)

Comment: From the reference to a hat given in this account, it appears fairly obvious that we have another case involving a "wild man." However, it is interesting to note that on the West Coast, some First Nations people refer to sasquatch as "gilyuk," which means, "the big man with the little hat." This description results from the creature's sagittal crest (pointed head) that from a distance appears to resemble a little hat.

PIKE COUNTY

Latham, 1999: On December 22, 1999 a young man and his girl friend had set up a burn barrel at the young man's grandmother's home near Latham. After the barrel was about one-third full of debris, they left it to go to the house for something. When they returned, they found that the barrel had been tipped over. It was too heavy for the wind to have moved it, so they thought that it tipped because it was not sitting straight. They went to look for a piece of pipe to place under the barrel. When about half way across the yard, they heard brush breaking in the surrounding woods. They then saw a large hairy creature staring at them at about a distance of about "two car lengths." The creature appeared to be about 7 feet (2.1m) tall. The young man hollered and the creature turned and walked away slowly, taking large paces and swinging its arms, "almost like a person, but definitely was not." The frightened couple then ran to their car. In recalling the incident, the young man remarked that the creature did not seem to be afraid of him and his girlfriend.

PORTAGE COUNTY

Nelson, 1977: On March 8, 1977 Mrs. Barbara Pistilli of Nelson telephoned the sheriff's office and stated that two teenagers had seen and shot at a giant furry creature. Deputies were told that the creature was eight feet tall (2.44m) with an estimated weight of 500 pounds (226.5kg). The creature had walked out of the woods and started across a field. It fled when shots were fired at it. Mrs. Pistilli stated that the creature had been seen before. (Source: *Record Courier,* March 10, 1977.)

PREBLE COUNTY

Rural Area, 1975: In July 1975 two 10-year-old children were playing on a farm in Preble County. They claim that they saw a strange creature watching them over the top of 6-foot (1.8-m) corn stalks. They ran and got a 12-year-old friend who also saw the creature. They tried to get the father of the older child to come and investigate, but he paid no attention to them. The children then climbed up on a shed roof and watched the creature. After a while, the creature ran off towards some woods and disappeared. The children described the creature as being upright, leaning forward and covered in long, smooth brown hair.

Eaton, 1976: A 1976 sighting at Eaton, is referenced. Footprints were found in the area (no details).

Roberts Covered Bridge, 1977: On May 18, 1977 two 13-year-old boys were walking their dog near U.S. 127 and Old Camdem Pike (vicinity of the Roberts Covered Bridge). The dog became very excited and got away from the boys, who immediately ran after it. Upon catching the dog, the boys detected a "rotten egg" odor. They turned around and claim that they saw a large hair-covered creature with very long arms and white eyes.

They estimated that the creature was about 9 feet (2.74m) tall and weighed about 500 pounds (226.5kg). The boys ran in fright and the creature chased them. They state that it was "right behind" because it took very large steps. When the boys reached the railroad tracks and the highway, the creature vanished from view.

Upon hearing this story, the mother of one of the boys notified the Preble County Sheriff's Department, and two deputies were sent to investigate the incident. While nothing was found, about one week later a farmer, who was one-half mile or so northeast of the encounter location, reported to police that there were two unusual footprints on his property. Bigfoot researcher Richard Hoffman was contacted and went to the scene. Hoffman states that the two prints showed five toes and measured 14 inches (35.6cm) by 7 inches

(17.8cm). The distance between the prints was roughly 6.5 feet (1.98m).

Eaton, 1977: In December 1977 another sighting at Eaton is referenced. Unusual odors were reported, and again footprints were found in the area (no further details).

Eaton, 1996: In response to reports of possible bigfoot activity in the Eaton area, researchers went to this area in April 1996. The details of their investigation follow.

Eaton Investigation, April 6, 1996: When the researchers checked with a local farmer for permission to go on his land, the farmer inquired if they were hunting the bigfoot that roams the woodlands from time to time. The farmer, in his sixties, revealed that when he was boy (9 or 10 years old), his grandfather told of strangers who frequented the area to hunt for the elusive "Grassman." Subsequently, several residents in the area asked to talk to the team. The team was taken to a home where about 20 people, ranging in age from toddlers to grandparents had assembled. They showed the team photographs of unusual tracks found in a cornfield, and told many stories of strange incidents – strange sounds, bad smells, feelings of not being alone, unusual animal behavior. Most related stories of a strange man-shaped beast that prowls the hills. No one present, however, had actually seen the creature.

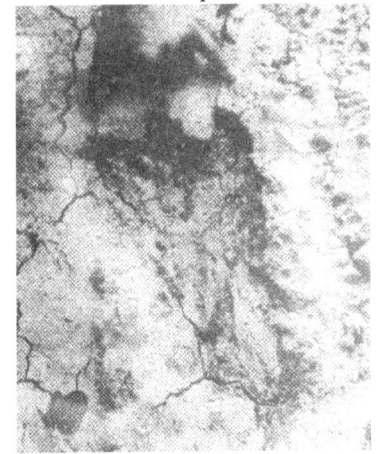
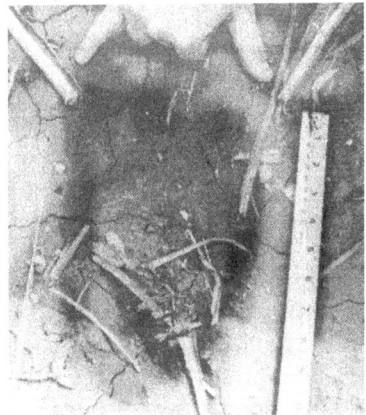
Three-toed prints, Eaton, 1996.

The group's investigation of the farmer's field and a wooded area revealed what appeared to be "markers" – geometric arrangement of sticks and sticks hooked on tree branches. Further, the configuration of breaks in a wire fence was unusual. It was also noticed that the bark on some trees had been peeled.

Possible three-toed prints about 9 to 10 inches (22.9 to 25.4cm) long were found, as shown here. Although not very clear, the second print shown was similar to a photographs taken by the residents. The group was very impressed with the people in this area. They were honest and genuine and stated the truth as they saw it.

Rural Area, 1996: On August 16, 1996 at about 9:15 p.m., a mother and her daughter saw what they believed was a bigfoot on the side of the road. The pair were heading east and after crossing a bridge, the car headlights illuminated an animal that the mother (driver) first assumed was a deer, although it appeared to have too much hair for a deer.

The mother stopped the car for fear that the animal would run across the road. One or more of the car windows were open and the two were overcome by a horrible stench. At this point, the car was about ten feet (3.05m) from the creature, which turned, looked at the vehicle, and then stood up on two legs with its arms extended. The mother noted that the creature's eyes were above the car's headlight beam. The creature then disappeared up a steep slope in three quick movements. Both mother and daughter were very impressed with how smoothly and effortlessly the huge creature moved. The two witnesses estimated its height at between 10 and 12 feet (3.1 and 3.7m). This sighting was reported to bigfoot researchers at the first opportunity and was investigated immediately. The details of the investigation follow:

Eaton Sighting Investigation, August 16 and 17, 1996: Bigfoot investigator George Clappison proceeded immediately to the scene of the mother and daughter sighting, arriving at 11:30 p.m. the same evening. Investigation revealed that there is a ditch on the roadside that is about 3 feet (91.4cm) deep. It is reasoned that the creature was crouched down in this ditch at the time the car approached. Given the mother's comment on the creature's eyes and the depth of this ditch, it may be reasoned that the creature was in excess of 10 feet (3.1m) in height.

In the exact spot where the sighting took place, it was noticed that the sparse ground cover was bent over. Further, the creature's path up the slope was devoid of spider webs that were highly evident in all other directions. Other investigators arrived on the scene the following night. On this night, one of the first things the group noticed was the level of noise made by insects in that area. Clappison then realized that this noise was not apparent the previous night.

On August 28, 1996 George Clappison revisited the scene of the sighting with other investigators. They noticed odors and other unusual conditions in the area. The details of their investigation follow:

Eaton Sighting Investigation, August 28, 1996: The team proceeded into the woods towards a creek during the early evening. In one spot they a noticed a strange arrangement of rotted logs. Two logs, about six inches (15.24cm) in diameter had been propped up against a tree at about a 15 degree angle. It appeared that the logs had actually rotted on the ground and were picked up and placed in the position described. A loud splash was heard, and then a series of splashes, like someone running through water. The splashes were followed by the sound of a tree crashing into the creek.

The investigators checked the area from which the sounds came, but did not find anything. The group had a heat scope, and as they withdrew from the creek, the scope indicated many heat sources – at one point, 7 to 10 sources. As they continued, the heat sources moved up along side of their location. At this time they sensed a repulsive odor that seemed to emanate from the direction of the heat sources. The odor intensified as they continued, and the heat scope indicated two heat sources moving along beside the team. It was now quite dark and one team member states that he momentarily spotted a dark figure moving in the gaps of some distant trees.

PUTNAM COUNTY

Pandora, 1954: In 1954 a sighting at Pandora is referenced (no details).

RICHLAND COUNTY

Mansfield, 1959: In March 1959 a 7-foot "thing" covered with gray hair and having large luminous eyes was reported near Mansfield.

Mansfield, 1963: In March 1963 there was a sighting of a creature similar to that reported in 1959 (Source: From a magazine article published in 1963, no details.)

Mansfield, 1973: In August 1973 a man living just north of Mansfield claims that he saw on his farm property an 8-foot (2.4-m) creature that was very wide and had long arms. The creature was standing by the man's barn at about 2:00 a.m. He shot at it with a shotgun, and the creature fled. He reported the incident to the local sheriff.

Butler, 1978: In late June 1978 residents of Butler reported to police that they had seen a mysterious creature in the area. Acting police chief Phil Stortz investigated the incident, but did not find any evidence of the creature. Stortz stated he had heard of half a dozen sightings that were not reported to the police. (Source: *The Columbus Dispatch, July 16, 1978 – Para-Hominoid Research, Dispatch State Service.)*

On July 8, 1978, two children, Eugene Kline and his sister, Kathy, who lived at 80 Elm Street, Butler, heard strange noises in a wooded area near their home. The pair subsequently claimed that they saw a creature about 5 feet (1.5m) away that they described as being 7 or 8 feet (2.1 or 2.4m) in height, with a head about 3 feet (91.4cm) in diameter, and having large red eyes. They did not recall whether the creature had fur or skin. The creature looked at the children in a strange manner while making an unusual growling noise. The children ran home and told their father, who called the police. The police searched the area; however, nothing was found. During the police investigation, the father told the police that his children said they had seen the same creature two weeks prior to this incident. No details on this first sighting were revealed.

Four days later, July 12, 1978, another one of the Kline children, Teresa, was badly frightened by two large red eyes staring at her in the darkness behind her home. Teresa ran into the house and went to her bedroom, where she was joined by her mother. The two stated they heard from the window a crying noise coming from the location where the eyes were seen. They also detected a strange odor which they claim came from the same location. The police were called and again searched the area with no results.

Reprints of the two police reports on the Kline sightings are as follows so the reader can see exactly what was reported to the police.

GENERAL OFFENSE AND ARREST REPORT, RICHLAND

COUNTY SHERIFF'S OFFICE, MANSFIELD, OHIO. JULY 8, 1978, WEDNESDAY 2300 HOURS. REPORTED BY COMPLAINANT: BIG ODD OBJECT: PLACE OF OCCURRENCE: AT THE SOUTH PORTION OF 80 ELM STREET, BUTLER, WORTHINGTON TWP, RICHLAND COUNTY, OHIO.

Upon arrival at the Kline residence located at 80 Elm Street, Butler, Ohio, I spoke to the complainant, Mr. Roger Kline, who stated at the above date and time his son *(and daughter)* (Eugene Kline and Kathy L. Kline) both who reside at 80 Elm Street were at the above place of occurrence when they suddenly heard a strange noise coming from the wooded area. Eugene Kline stated at this time he looked up and saw directly in front of him approximately 5 feet away an unknown object believed to be approximately 7 or 8 feet in height, unknown weight, having a head approximately 3 feet in diameter with very large red eyes. Eugene Kline along with his sister Kathy stated that they both did not know whether this object had fur or a skin. They stated the subject looked at them very strangely making a very unusual growling noise. They stated at this time they turned around running back to their residence on Elm Street, advising their father (Roger) of the situation. Mr. Kline at this time stated that he does believe his daughter and son due to the fact that approximately two weeks ago prior to this occurrence they had seen the same strange object. He stated himself he did go back in the area with negative results of finding this object. He does believe there is something back there without knowing what it is exactly, what it is that is frightening his children. At this time I advised the complainant along with his son (Eugene Kline) that this officer would walk back to the area where the object was seen to attempt to locate or find any clues or evidence of this object being there. A check of the area was made with negative results of seeing or hearing any strange or unusual things. At this time Mr. Kline stated that during the daylight hours on 7-9-78 that he was going to make a check of the area and would advise this officer.

GENERAL OFFENSE AND ARREST REPORT, RICHLAND

COUNTY SHERIFF'S OFFICE, MANSFIELD, OHIO. JULY 12, 1978, WEDNESDAY 2305 HOURS. REPORTED BY COMPLAINANT: SAW OBJECT: PLACE OF OCCURRENCE: ABOUT 300 YARDS SOUTH WEST OF 80 ELM STREET, BUTLER, WORTHINGTON TWP., RICHLAND COUNTY, OHIO.

Upon arrival at the Kline residence located at 80 Elm Street, Butler, I spoke with complainant Roger Kline who stated that on the above date and times, his daughter Teresa was outside the residence in their yard unloading hay with him, Eugene Kline, Kathy Kline. Mr. Kline stated that at the above date and time his daughter had saw the unknown strange object standing at the above place of occurrence. At this time I spoke with Teresa Kline in reference to this matter and Teresa stated that at 2330 hours she was helping her father unloading some hay at which time she heard the train going down the tracks located behind their residence and she stated that this is not unusual that it does pass through approximately 4 times daily. She stated that the unusual thing about this incident is that train kept blowing its horn repeatedly for an unusual length of time. She stated that at this time she turned around in the direction of the train with her 6 cell flash light turned in that direction and at which time she did observe these 2 large red eyes believed to be the size of golf balls staring at their residence from the place of occurrence. She stated that she could not see the shape of the body, all that she observed was the eyes. She stated that they were orange-red in color and Teresa said they looked like the end of a light cigarette, that sort of color. She stated that at this time that she did scream and throw the flashlight and run into the house, stated that everyone else entered the house with her. She stated that she did run upstairs crying to her bedroom. Teresa said that she heard a crying noise outside of her residence like the sound of a real cat but in a deeper tone of voice, real loud coming from the place of occurrence. Kathy along with her mother stated that there was a funny smell that they observed from the bedroom window. They believed that the scent that they smelled was coming from this object. They further stated that it smelled very similar to cow shit. At this time myself, Roger Kline and Eugene Kline did go to the place of occurrence searching the complete immediate area where this object was supposedly seen with negative results of finding any clues or evidence of anything having been in the area. At this time we returned to the house I further advised Mr. Kline to have all of his children stay in the yard and not go back to the area where this object was seen. At this time I left the Kline residence meeting with Sgt. Frontz along with Chief Fred Horn of the Butler Police Department. At 2350 hours 7-12 this officer along with the other two mentioned did go to S Road, the last residence on the west side before you come to the bridge. Chief Horn stated that the house was vacant and he would like this officer along with Sergeant Frontz to go there to check the vacant building. Upon arrival a complete check of the building was made with negative results of finding anyone around or any clues pertaining to the case.

When speaking to Teresa Kline earlier in reference to this, she stated that she was the only one in her family that saw this object at the above date and time.

At this time headquarters was notified of the negative results and patrol was resumed.

Butler/Bellville Area, 1988: In mid-August to early September 1988 two animal trapper hobbyists (both in their 50s) were setting traps in the forest between Butler and Bellville. Over a three to four week period, they heard piercing cries in the forest. They were impressed with the volume and intensity of the cries, which they could not recognize as those of a known animal. Sometimes the cries were almost musical, somewhat like whistling.

During their travels, they found large human-like footprints in soft soil heading up a hillside. The prints were about 15 inches (38.1cm) long and showed "huge strides." On a quiet afternoon, they observed a strange creature at about 100 feet (30.5m). It had medium length brown hair all over its body except for its face, a low forehead, small dark eyes, huge hands, a bulky build and a "nasty smell." The creature looked at the men for a few seconds and then "took off" up the hillside. The men collected their traps and have not returned to the area since the sighting.

SCIOTO COUNTY

Sciotoville, 1930s: In the early 1930s George Johnson, about ten years old at that time, claims he saw a coal-black hairy creature, like an overgrown gorilla, walking upright in the woods. He estimated the height of the creature at 8 feet (2.4m) and felt its weight was about 700 pounds (317kg). He noted that the creature was very muscular. He could not see any facial features. (*Newsletter of Intriguing Studies,* 1981.)

Shawnee State Park, 2003: On June 18, 2003 a man and his wife found large human-like footprints in Shawnee State Park. They reported the find to a bigfoot "hotline" in California, and Joedy Cook of the Ohio Center for Bigfoot Studies was subsequently contacted. Cook investigated the finding and made a plaster cast (seen here) of one of the prints which measured about 15 inches (38.1cm) long.

Cast of footprint found in Shawnee State Park. The well-defined toes make this one of the best casts taken to date.

STARK COUNTY

Alliance, 1957: In 1957 a bigfoot sighting at Alliance is referenced (no details).

Massilon, 1973: In August 1973 a sighting at Massilon is referenced (no details). In October 1973 a creature 7 feet (2.1m) in height and with a strong odor was reported in the Massilon area. The police were notified. (Source: *Akron Beacon Journal,* October 27, 1973.)

Paris Township, 1978: On the evening of August 21, 1978 Mr. and Mrs. Herbert Cayton and four friends, all residents of Paris

Township, claim they saw a 6-foot (1.8-m) creature covered with hair sitting atop a backyard chicken coop at the Caytons' residence. One of the witnesses drove a car across the backyard with its headlights on to get a better look at the creature. When the creature ran towards the car, the five other witnesses ran into the Cayton house. The creature then began looking in a window at them. Facial features could not be seen because of the creature's bushy hair. The group also heard footsteps on the roof. When Mrs. Cayton loaded a gun, the creature disappeared.

The sheriff was called and when he arrived he noted that the people were visibly shaken, some were even afraid to go to sleep. There were two footprints found – one distinguishable, the other not as good. The sheriff scouted the heavily wooded area, but found no sign of the creature.

Mr. and Mrs. Cayton said they had seen the creature several times over the past few months. Four other residents said they had seen the creature and two smaller similar creatures on a previous occasion. (Source: *The Cleveland Plain Dealer,* August 24, 1978.)

This incident was thoroughly investigated by Ron Schaffner and other bigfoot investigators. The details of the investigation follow:

Paris Township Sighting Investigation, August 1978: It was revealed to investigators that when the Caytons and friends heard noises in the direction of the old chicken coop, they saw two pairs of yellow eyes reflecting the porch light. When the car headlights were turned on to get a better look, the eyes appeared to be those of two "cougar-type" felines. The party then saw what looked like a large bipedal hairy creature step in front of the large cats as if to protect them. The creature then proceeded to lurch towards the car.

Subsequent investigation revealed that Mrs. Cayton first observed the creature about three weeks prior to the chicken coop incident. At this time, the Caytons' grandchildren came running into the house in a frightened state claiming they had seen a large hairy monster in a pit near the house. The bush around this pit had been recently cut down by Mr. Cayton, and some garbage had been dumped in it for the raccoons. Mrs. Cayton, Howe Cayton and Mrs. Keck went out to the pit and claim they saw a creature that was covered in dark matted hair. They estimated that it was about 7 feet (2.1m) tall and weighed about 300 pounds (135.9kg).

Mrs. Cayton stated that she saw the creature again in the pit some time later. She stated that it was sitting in the pit picking at the garbage. She could not see its features due to the amount of long hair covering its face. However, she recalled the creature had no visible neck.

More sightings in the same area of the same or a similar creature were reported to the investigators as follows:

On August 22, 1978 Mrs. Mary Ackerman of Minerva drove to the Caytons' residence. As she turned into the driveway, she claims she saw the creature previously seen by the Caytons standing on top of a hill next to the local strip mine. She watched it until it walked out of view.

On August 23, 1978 an unusual creature was reported again at the Caytons' residence. Howe Cayton, who was not sure if it was the same type of creature as previously observed on August 21, 1978, fired a gunshot into the air and the figure departed.

On August 26, 1978, the alleged creature was reported in the same Paris Township area by John Nutter and his brother, Jerry. John Nutter said that at 2:00 p.m. on that day, he and his brother went into the wooded area near where the creature had been previously sighted. He said that as they walked toward an abandoned strip mine pit, they heard noises like a person or animal walking parallel to them, hidden by the dense brush. Nutter said as they approached the pit, a large, hairy creature stepped out from behind a tree and emitted a loud scream which he described as a cross between an elephant and a trumpet or trombone.

At this point, Nutter snapped one picture of the creature through a telephoto lens before he and his brother began to run. He said he looked back once and saw the creature running almost parallel to him on two legs. Nutter said that in his panic he screamed, "it's a bear," as he emerged from the woods. He said he knew subconsciously it was not a bear, but had no other logical explanation.

The Nutters called the sheriff's department and deputy Robert Sylvester, assigned to the wildlife department, went back into the woods with them at about 4:30 p.m. They found several large footprints that have not been identified. While they searched the woods a second time, they heard the same strange cry previously described coming from some dense brush.

John Nutter said he has seen hundreds of bears near his previous home in West Virginia, and would not have run if the creature he saw had been a bear. He said he could not see any eyes or ears on the creature, which was about 50 feet (15.2m) away. Also, he said bears cannot run on two legs as this creature did.

The brothers described the creature as about 5 feet (1.5m) tall and weighing more than 300 pounds (135.9kg). They said it had dark brown or black shaggy hair covering its body and face. John Nutter said the birds in the woods became silent just before the creature was sighted, but he did not sense an odor. When the photograph was developed, all that appeared were trees and brush.

On September 8, 1978 Mrs. Mary Ackerman claims she observed two ape-like creatures near the strip mine at Paris Township. She watched them until they disappeared in the thick weeds.

Paris Township, 1979: In 1979 Herbert Burke Jr., stated he saw a bigfoot while driving near the town trailer park where he resides. The creature was crossing Route 30 at night. Burke positioned his car headlights directly on the creature, which was about 40 yards (36.6m) away. He described the creature as 7 or 8 feet (2.1 to 2.4m) tall, weighing more than 400 pounds (181.2kg), and covered with dark matted hair. Burke also mentioned that other trailer park residents often heard rocks hitting their trailers at night, together with a variety of strange noises coming from the woods.

Paris Township, 1989: In June 1989 the *Akron Beacon Journal* did a follow-up on the August 1978 Paris Township sightings. The reporter talked to Mrs. Cayton, who stated that she often finds large footprints in the area. She said they range from 14 inches to 21 inches (35.6cm to 53.3cm) long. She went on to state that, upon the advice of Deputy Shannon, she continually puts fruit and vegetables out on a table behind her place in the evening. In the morning the food is gone, and occasionally large footprints are found nearby. Further, sometimes Mrs. Cayton and others experience the same eerie silence and smell the odor that occurred in 1978. These conditions, they felt, indicate that the creature is nearby. (Source: *Akron Beacon Journal*, June 29, 1980.)

Alliance, 1991: In December, 1991, Ron Brunner an Alliance farmer, claims he saw a 9 foot (2.74m) tall creature in a field across from his house. He reported that his cows were very frightened. (Source: *The Wall Street Journal*, February 24, 1992.)

Forested Area, 1992. In November 1992, a bow-hunter observed an upright walking creature apparently stalking deer. See Chapter 7, **Ohio Deer Kills and Bigfoot**, for details.

SUMMIT COUNTY

Cuyahoga Falls, 1979: On January 9, 1979 a sighting at Cuyahoga Falls is referenced (no details).

Akron, 1988: In 1988 a sighting at Akron is referenced (no details).

Barberton late 1980s or early 1990s: During this period resident Barbara Bilinovich claims that three "humongous" creatures chased her from the woods one night (no details).

Akron, 1995: Two Akron residents reported unusual incidents, including a possible bigfoot sighting, in the Kenmore area over a number of years. Researchers visited the residents in February 1995 and accompanied them to the location of the incidents. Many anomalies were observed. The entire sequence of events and details of the investigation follows.

NOTE: The two witnesses, for the purpose of this account, are named Dale and Tim Atkins, as they did not wish to have their actual names published. Dale was 43 years old at the time and is the father of Tim.

AKRON Sighting Investigation, February 19, 1995: On February 19, 1995 Joedy Cook, George Clappison and Terry Endres traveled to Akron, Ohio to interview Dale and Tim Atkins on their alleged experiences with a possible bigfoot in the Akron area. The day was mostly sunny with a slight haze; the temperature was about 50 degrees Fahrenheit. The ground had only spotty snow remaining, and the soil was frozen, except in places where the sun had thawed the top layer.

Dale Atkins and his son live in the Fairlawn area of Akron. The team arrived at Dale's small white house at 12:30 p.m., and after introductions and general planning, proceeded to load their gear into Dale's van for travel to the location of the occurrences. It was a one-hour drive to this location, which was in the Kenmore area (Dale's boyhood home) between Manchester Road and Main Street. Kenmore is a suburb of Akron.

The interview with Dale revealed that he and his boyhood friends had been aware of an unusual creature in the Kenmore area since Dale was about 13 years old. He had spent a lot of time in the woods during his boyhood, camping and fishing with friends, and the creature had made itself known on a number of occasions. Mostly, it was heard crashing through the woods or splashing in the swamp. The boys found unusual three-toed tracks a number of times that they attributed to the creature.

Dale continued to camp and fish in the area throughout his adult life. When his son, Tim, was old enough, he accompanied his father. Dale mentioned that adult friends, who also used this area, had told him stories of their "run-ins" with an unusual creature.

HISTORY

A local radio show in 1995 prompted Dale and Tim to share their unusual experiences. The radio show featured a bigfoot researcher talking about bigfoot creatures in Ohio and elsewhere. Dale called the radio station, but was unable to talk to the researcher. However, the station sent Dale a form to fill out on his and Tim's experiences. The form was submitted, and Dale and Tim were featured on a later (1995) radio program. A copy of the form found its way to the Ohio Bigfoot Research & Study Group. Consequently, this group became involved in the case.

In time, Tim followed in his father's footsteps and frequented the area with his own friends. One evening, Tim and a group of boys went night-fishing. When they were some distance into the woods, they stopped on a wooden bridge to decide on the best route to the fishing spot, and to check if they had enough bait with them. By this time it was dark. They were all standing on the bridge more or less facing each other. Without a sound, what the boys describe as a "large dark hand" seemed to come out of the darkness and rested gently on the shoulder of the boy carrying the fishing poles. Utter chaos followed, and all fled leaving poles, tackle and bait on the spot. They returned the next day and found their gear as they had left it. The walkway of the bridge where the boys were standing was about eight feet (2.44m) above ground level.

Comment: We won't even speculate on how tall the creature had to be to reach the boy's shoulder from ground level. It is certainly more comfortable to write-off this story as the product of over-active imaginations.

Another very unusual incident occurred in 1988 while the family was still living in Kenmore. Tim was alone in the area and decided to inspect one of the caverns formed by the build-up of construction debris dumped in the area over the years. He noted that the entrance to the cavern was worn smooth. As he was about to enter, a large rock, about the size of a softball, thudded down near his knee. Tim estimated the rock was thrown at him from a very long distance – possibly 100 yards (91m). The rock dropped straight down and did not roll when it hit the ground. Tim ran home and reported this occurrence to his father, who went back with his son to investigate. After they arrived at the cavern location, another rock fell, followed in time by more rocks. It was now dark so the two made a hasty retreat. Rocks continued to fall near them right up to the point where they cut through a neighbor's yard to get back to their home.

During the team's investigation, both Dale and Tim referred to the creature as the *Grassman*. From their experiences, they said it seemed playful. While unable to provide any real details, both said they had glimpsed what they believed was the creature.

The site investigation revealed that the area frequented by Dale and his son consisted of a cleared area and a wooded lot that adjoined a fairly large section of forested land. The wooded lot, in which the encounters took place, is bordered by the Ohio Canal on the east side and a junkyard on the west side. In one area, it could be seen that in years past a part of the land had been a swamp. Over the last 15 years or so, the cleared area leading up to the wooded lot has been used as a dumping ground for road construction waste and other commercial debris. Piles of bricks, large slabs of concrete, old gas tanks with their ends cut off, piles of rotten lumber, and an assortment of other junk litters the area.

Tim informed the group that the area in which they were now located was private property belonging to the junkyard owner. Tim stated that he knew the owner, and was sure that he would not object to the presence of the group. The team's first impressions were that the wooded lot was not large enough in itself to support a bigfoot type creature. However, as the lot had passage to a larger forested area, support for a part of a year seemed possible if the creature migrated into the larger area.

This last statement is made with some reservations as there is light to heavy development beyond the larger forested area and in all other directions – hardly ideal for a bigfoot. It is possible that passage to Long Lake (east side of sighting location) could be made at night through a network of lightly developed or undeveloped pockets. The lake, however, would limit access to forested regions on its northeast shore. Nevertheless, even if the creature could overcome this obstacle, further progress east would be greatly hampered by development.

As Dale and Tim had experienced an on-going presence of the creature at this site, the investigators conducted a thorough search of the area. They found impressions in the snow that looked like they may have been melted-out footprints, and other impressions in the ground that could also have been footprints.

Upon making their way through an open field in waist-high growth to a wooded area, they found an unusual nest-like construction of sticks and weeds, as seen here. It measured about ten feet (3.05m) long by three feet (91.4cm) wide. The bottom layer consisted of up to one inch (2.54cm) diameter sticks piled 6 inches (15.2cm) or so high. This arrangement was covered with a layer of plucked foliage, grass, weeds and twigs. The nest appeared to be about one year old.

There was an electrical tower in the area with high bushes at its base. The bushes were bent down in some places, indicating they had been disturbed by someone or something. What appeared to be a three-toed footprint, 10 to 12 inches (25.4 to 30.5cm) long and 5 inches (12.7cm) wide at the ball of the foot, was discovered nearby. The heel was narrow, like that of a human. The big toe seemed offset as though the foot was deformed. The team concluded that this was probably a handprint rather than a footprint.

Close-up photograph of the interior of the nest. A plastic bottle is seen about center in the background.

As the team proceeded further into the wooded area, they found a domed structure made of forest material. The following photograph shows the structure being inspected George Clappison.

The structure was hollow on the inside with enough room to accommodate three men in a seated position. There was a pungent odor inside and hairs were found on the earth floor. The structure appears to have been made by using large branches to form a tunnel. Then, smaller branches were placed on top of the larger branches with vines and weeds woven-in. This assembly was then covered by other forest debris, and finally a total covering of long grass. In many respects, the structure resembled an igloo. While it would provide effective protection against the elements "as is," a covering of snow would greatly increase its effectiveness.

What appears to be another handprint was found in mud near the structure, and further investigation revealed a number of dug-out holes. They were about 6 inches (15.2cm) wide and 12 inches (30.5cm) or so deep. The dirt was piled-up to the side of the holes. Such holes found in other areas have been associated with animals (perhaps bigfoot) digging for roots. It was also noted that sumac bushes north of the electrical tower were missing berries on their upper parts – 8 to 10 feet (2.4 to 3.1m) up. The branches appeared as though the berries had been "skinned off" (i.e., pulled through one's hand or mouth). The sumac bushes were spaced, between 2 and 3 feet (61 and 91.4cm) from each other. This is likely a natural spacing as a result of re-seeding. Curiously, it appeared as though one of the plants had been removed. A hole 2 feet (61cm) wide and 1 foot (30.5cm) deep was found where the plant would normally have been. There was no excess soil around the hole and no up-routed plant was found in the area. The indications are that the plant had been manually pulled out of the ground. As the sumac bush is fairly tight-rooted, manual removal of the plant without mechanical aid would be highly difficult, if not impossible.

Other domed nest structures were later found in another county (see sidebar). A television program showed the Kenmore structure, and they all appeared in at least one magazine article. That they might be associated with bigfoot is, of course, possible, but the fact that they have been found only in Ohio greatly restricts the probability. A full analysis on the structures is provided at the end of this chapter.

SIMILAR STRUCTURES

Two similar structures, shown here, were later found in a remote wilderness area of Hamilton County, They were about 200 yards (183m) apart. and located about 100 yards (91m) from the Indiana border. George Clappison is again seen inspecting the oddities. The team referred to these structures as a possible bigfoot "nests." However, I believe they would be more appropriately called "hollows."

TRUMBULL COUNTY

Hubbard, 1997: In January 1997, a schoolteacher was driving along Route 304 near Hubbard at about 6:00 p.m. There was snow on the ground and light snow was falling. The teacher saw a black hairy ape-like creature cross the road. She stopped her car and reached for her camera. Unfortunately, the creature was out of view before she had time to get a picture. She did, however, take a photograph of one of the creature's footprints in the snow. Being a teacher, she had a ruler with her and laid it beside the footprint to give an indication of the print's size. The photograph she took is shown here on the left. On the right is a cast (taken by Bob Titmus) of a footprint found in California in 1958 that is very similar in shape.

Comment: If a cast were made of the Ohio print, it would be very close to 15 inches (40.6cm) long. I have considered the possibility that the Ohio print could have been made from the California cast, copies of which were available for purchase in the mid 1990s. However, a prints made from this cast would be slightly larger than the cast (up to 15.5 inches/39.4cm). It is possible a cast was obtained and then sanded down at the heel to shorten it, but I really don't think this was the case.

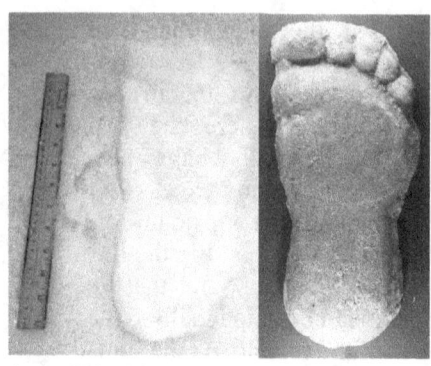

Left, Ohio print which measures about 14.5 inches (36.8cm) long.
Right, California cast, 15 inches (40.6cm) long. Note the contour line on the left of each illustration.

TUSCARAWAS COUNTY

Newcomerstown, 1992: In August 1992 and April 1995, sightings at Newcomerstown are referenced (no details).

UNION COUNTY

Rural Area, 1980: On June 17 1980 Patrick Poling, a farmer, states he saw a monster on this farm that was about 7 feet (2.1m) tall and could weigh 400 pounds (181.2kg). Poling was plowing a field that evening before dusk when the beast burst out of the woods and lumbered along the edge of the clearing. He watched the creature until it was about 100 feet (30.5m) from him. The creature then turned, looked at Poling and ran back into the woods. Poling states that the creature looked like a big hairy ape that walked like a man. He said it had long black hair that wasn't fur because it hung straight down. He rushed to a neighbor's house to report the amazing incident.

The next day, Poling and a few neighbors went in search of the beast. Near the edge of the woods, three huge four-toed footprints were found. A plaster cast was made of one of the prints. It measures

17 inches (43.2cm) long, 7 inches (17.8cm) wide, and was 2 inches (5.1cm) deep. Officials at the Columbus Zoo in Ohio were provided with the cast for study. (Source: *The New York Globe,* June, 1980.*)*

On June 24, 1980 Donna Riegler of Marysville states she was returning home from work when she saw a gigantic, hairy creature lying on the highway. She was so frightened, she put her car into reverse and backed away from the beast. Riegler reported it was covered with hair except for the palms of its hands, and that it stumbled away with a robot-like walk. (Source: *The New York Globe,* June, 1980.*)*

Comment: Both of these incidents were reported to the Union County Sheriff's Department and officers were sent to investigate them. The officers stated there was no doubt in their minds that somebody saw something "out there."

Marysville, 1980: In late 1980, a sighting at Marysville is referenced (no details).

Rural Area, 1985: On June 19, 1985, Claudia Beeson was cycling on a road near a Union County creek. She claims that she saw a bigfoot running close to her. Beeson stated that the creature looked like a boxer doing road work. (Source: *Cleveland Plain Dealer,* August 9, 1987.)

VAN WERT COUNTY

Ohio City, 1981: In this year 15-inch (38.1-cm) footprints were found near some seldom-used (about once a year) railroad tracks in Ohio City. The Mammal Research Team of Lima, Ohio investigated the finding.

The team followed the footprints which led to a local pig farm, and then back to the railroad tracks. As the team members proceeded down the railroad track they found the head of a baby pig. They then found 15 dead baby pigs, still with umbilical cords attached, at intervals along the track. The team concluded that a bigfoot had gone over to the pig farm around farrowing time and scooped up 16 newborn pigs. The bigfoot placed the piglets in a feed sack, and as it walked down the tracks it ate one piglet, all but the head. Then the sack got wet and the piglets began to fall out. The report does not mention the team finding a discarded feed sack; however, it is apparent one was found.

About a mile (1.6km) away from this find, the team discovered about 30 to 40 feeder pigs that had been lying there about 6 months to a year. They too were in a feed sack. We have no explanation as to why the bigfoot apparently abandoned his catch.

VINTON COUNTY

Rural Area, 1970s: In the 1970s, a sighting at Vinton County is referenced (no details).

McArthur State Park, 1980: During 1980 it was alleged that a youth shot and injured a bigfoot in McArthur State Park. The creature became very aggressive, threw large rocks at house trailers, and then approached the units and hammered on the sides of them. It was reported that the creature tipped-over some of the trailers. People in the area quickly left because of the creature's activities.

Forested Area, 1980: On August 24, 1980, Larry E. Cottrill claims he saw three bigfoot-like creatures close to his home, which is near McArthur. He shot at them and may have wounded one in the shoulder. (Source *Cleveland Press*, October 6, 1980.)

Also on August 24, 1980, two Vinton County hunters claim they saw a pair of bigfoot at a distance from the back (i.e., back view). One of the hunters fired his gun in the direction of the creatures. After the shot was fired, the hunters saw one of the bigfoot drop its right shoulder. It was speculated that the creature may have been wounded.

This report was provided by bigfoot investigator Bob Gardiner, who further stated he and his team have found many bigfoot footprints in the Vinton area, along with a possible handprint. Casts of two footprints and a hand cast are shown here. In all, they found six different print sizes. Gardiner notes that the weights of the creatures

Casts of footprints and a handprint found in Vinton County.

appear to vary as the prints had different ground depths. The smallest footprint measured 8 inches (20.3cm) long and the largest was 17.5 inches (44.5cm) long. Gardiner thinks the creatures shot at stayed close to Vinton County because one of them could be wounded and may have had an infection. (Source: *The Daily Olympian,* October 9, 1980.)

McArthur (Town), 1980: On October 11, 1980, Rodney Peoples states he saw a bigfoot in McArthur at about 7:00 p.m. He said it was in a power line right-of-way clearing, about 150 yards (137.2m) from his home. (Source: *Columbus Citizen Journal,* October 14, 1980.)

SALT FORK STATE PARK

Forest Region, 2004: On August 18, 2004 a couple (man and woman) from Cambridge, Ohio claim they heard and saw a creature with bigfoot characteristics near a groomed trail in Salt Fort State Park. The incident was investigated on August 22, 2004 by Marc DeWerth and two fellow investigators with the Bigfoot Field Researchers Organization. Possible "stick stacks" were found (an oddity associated with some sightings in which sticks have been placed around a standing tree), along with a possible footprint about 9 inches (22.9cm) long.

Comment: Salt Fork State Park has had an active history of bigfoot sightings. We are told there were six or seven sightings in the first two weeks the park was opened. Other incidents followed; however, we do not have specifics. The tree carvings seen here, found on a dying tree in the park, are said to represent a "Woolly-Booger," another name for bigfoot or sasquatch. The photograph was taken by Robert Morgan in 1999. An elderly gentlemen who had once lived in the area informed Morgan what the carvings represented. He further stated that, as a child, he had seen the creatures on many occasions.

Tree carvings of a "Woolly-Booger."

WAYNE NATIONAL FOREST

South Forest Region, 1966: In 1966 a 7-foot (2.1-m) "ape man" was reported roaming in the southern tip of the forest.

General Forest Region, 1980: In August, 1980, a sighting is referenced. Footprints were found (no details).

In October 1980, bigfoot investigators Bob Gardiner and Dean Cottrill reported that they found a possible bigfoot footprint and possible bigfoot blood in the forest. The blood was found on a tree. They also believe they heard the creature in this area. They stated that subsequent analysis of the blood indicated that it came from some sort of primate. (Source: *Columbus Citizen Journal,* October 9, 1980.)

Eastern Forest Region, 1994 and 1995: In each of these years, a forest ranger found large human-like footprints and made casts of the prints, as seen here. The cast on the left is about 17 inches (43.2cm) long, and the cast on the right is about 16.5 inches (41.9cm) long. The similarity of the cast on the right to that of a cast taken on Blue Creek Mountain, California by Peter Byrne in May 1960, is very evident. The following illustration shows a comparison.

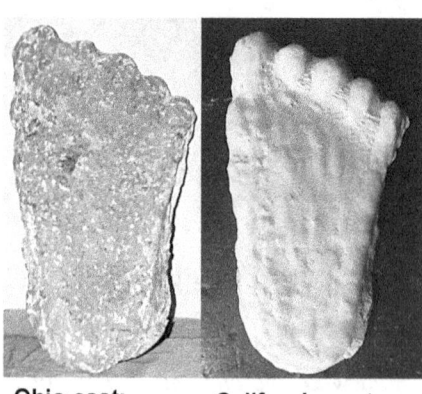

Ohio cast:
16.5 in. long
(41.9cm)

California cast:
14.5 in. long
(36.8cm)

Comparison of Ohio Cast and California Cast

Both casts are shown here at about the same length for comparison purposes. While the casts are different lengths, their similarity is remarkable.

ANALYSIS OF THE OHIO DOMED "NEST" STRUCTURES

The possible Ohio bigfoot nest structures present a lot of questions. If they are not associated with bigfoot, with what are they associated?

One possibility is that they are natural structures caused by wind. In other words, high winds picked up forest material and deposited it in a pile. However, the structures consist of larger branches covered with smaller branches, and then covered with grass. Wind would not make these distinctions in forming the structures. We also have a problem with the hollow on the inside – how could this have been made by the wind?

Another possibility is that the structures were made by children as "forts," which would account for the intelligence factor in the actual construction, including the hollow on the inside. While this explanation is certainly possible, the "forts" are not of the type generally made by children – they are too primitive. Children are a lot more inventive in this regard, and usually make a lean-to or a comparative structure. We also have the problem that "nests" or "hollows" have been found in two wide-spread areas. It is not likely the same children made all of the structures. Could different children have made the structures? In this case, we would have to say that the structures are of the type commonly made by the children in Ohio. However, if this were correct, the investigators would certainly have known about it, and would have disregarded the structures in the first place.

Another suggestion is that the structures are the result of a bush clearing crew working in the area. In other words, the crew deposited the material in a pile for subsequent pick-up or burning. Again, we have the problem with the seemingly planned placement of debris that resulted in a hollow on the inside. Further, the investigators did not notice any other clearing crew activities in each area, Nevertheless, while this *might* be a consideration for the Kenmore structure, those in Hamilton County were too remotely located for this kind of activity. It might also me noted that the Kenmore structure was on private property, so a government clearing crew was definitely not involved.

We next come to the possibility that the structures were made by some known animal. It has been offered that raccoons are known to make structures of this nature. Here, we have a problem with the size of the structures They are far too big to have been made by a raccoon, or many raccoons. Whatever the animal was, it had to be fairly large. Coincidentally, a primatologist in California stated that gorillas have been known to make similar structures. We know there are no gorillas in North America, notwithstanding bigfoot or Ohio's *Grassman,* (certainly the same creature).

We must also come to grips with the question of *locality* for the Kenmore structure. It was found in an area that is virtually "urban-locked." In other words, the area, while itself is undeveloped, is completely surrounded by development. Indeed, an Akron resident, John Sawvel, pointed out to me that the Kenmore structure was very near a large building.

As pointed out, there is a small wooded lot adjacent to the area that extends into a larger wooded section. However, the woods are eventually terminated by development. Surface passage to this area from remote wilderness areas requires travel through regions that are highly residential or commercial. It appears unlikely that the structure "builder," whatever it was, could surface travel to larger wilderness areas without being seen fairly often. It is equally unlikely that it resided in the area for an extended time period because the area itself does not appear large enough to support it. In other words, it would be severely limited in its food sources.

As with most issues associated with bigfoot, we come to a dead end on the structures. Certainly, they may not be even remotely connected with bigfoot. Nevertheless, they are now fully connected to Ohio's bigfoot "lore," and I must admit that do appear to be a very appropriate for the creature.

Update: Real or not, Grassman appears to still be alive and well in Kenmore. In January 2001, the *Akron Beacon Journal* included a reference to the creature in an article on Kenmore. Headed, ***Bigfoot-like creature, Grassman, haunts area,*** the article states: "For years some Kenmore residents have reported seeing an oversized, hairy, ape-like creature wandering in wooded areas. They called it Grassman for reasons now obscured by time." The article then goes on to mention the radio show with Dale and Tim Atkins, together with findings by the Ohio Bigfoot Research & Study Group.

Chapter 5
Sighting Considerations

Sighting Credibility: The credibility of sightings is mainly supported by the fact that many different people see a similar creature, and there is physical evidence found in the form of footprints. The prominent researcher John Green has pointed out that if just one report or one print is proven to be genuine, then we have a solid case for bigfoot. While one would think that the number of bigfoot incidents would have resulted in some conclusive (non-contestable) evidence supporting bigfoot existence, such has not surfaced.

However, even if we take a highly rational and "scientific" approach to the bigfoot issue and assume people are "seeing things," we cannot ignore the footprints. There are just too many of them in too many isolated locations to assume they are all faked. When incidents involve both a sighting and footprints, the report certainly takes on a higher level of credibility. If the witness was "seeing things," what left the footprints?

While fakery (person in a costume) is possible in some cases, or even many cases, such does not appear *probable*. In short, there does not seem to be *reasonable grounds* for someone to perpetrate the hoax. Moreover, there is also a very high *risk factor* in a hoax of this nature. People who have guns and feel they or their property may be in jeopardy tend to "shoot first and ask questions later." Not only that, but many people in rural areas have dogs. While these animals might be afraid of a bigfoot creature, they would not have the same fear for a hoaxer. There are certainly many other considerations (trespassing, accidents on unfamiliar ground and so on), but what we have provided appear to be the most crucial factors.

The nagging question on sightings is why don't we have more photographic evidence? This question is particularly frustrating if the witness had a camera (or immediate access to same) when the sighting took place. It appears that people often become so pre-occupied or excited that they are unable to "get themselves together" when the incident occurs. In other cases, the moment is so brief they don't have time to position and focus their camera. There is yet another reason. When I asked some First Nations youths (who firmly believed in bigfoot) in Bella Coola, British Columbia the question relative to getting photographs, they simple responded, "Why?" In other words, "I don't need to be convinced, if you do, that's another story." Money was suggested as an incentive, but this did not seem to impress the group.

It may be, however, that there are good photographs, but people are reluctant to make them known. Indeed, this fact might be true for not only photographs but other hard evidence. The only real incentives for people to come forward with such material is notoriety and perhaps financial gain. Many people are not interested in these "rewards." It appears their satisfaction in harboring the knowledge takes precedence, or they do not wish to become the center of attention on such a controversial issue.

Probable Causes of Ohio Sightings: Once we can, as it were, "get over the hump" regarding the probability of a bigfoot creature, we can start considering other questions such as why sightings are taking place. Specifically, why is this creature "invading" our (human) territory. Granted we have invaded its territory, however, most sightings take place in our territory. We have reasoned that there are enough forested regions for it to exist, so why does it show up in semi-rural or rural communities, and is some cases urban communities? We know that the creature is very human-shy, so why does it risk human contact? The answer appears to be that the creature is looking for "easy food," or perhaps more appealing food than it finds in the wild. It might be noted that on the Pacific coast, it has been reasoned that there is more than an abundant supply of naturally available food for these creatures. There continues to be, however, numerous sightings near human populated areas or areas in which humans are often found (recreational areas).

Given Ohio has the same food resource level, it appears that if the Ohio creatures wanted to remain totally concealed, they could do so, save rare sightings by people who go deeply into forested areas. It is noted, however, that some incidents indicate the creature is highly curious, bringing into play another factor that results in sightings.

Night Sightings: About half of all bigfoot sightings occur at night. Those that occur at this time are mostly in urban areas. Bigfoot, however, is not considered to be a nocturnal creature. Bigfoot has undoubtedly reasoned that night is the best time to trespass on man's domain. The reasons are obvious – there is less chance of being seen and there are fewer people around to do the seeing. Night raids are a natural process for nocturnal creatures, for bigfoot they appear calculated.

Sighting Frequency: Bigfoot, of course, because of its height and size, runs a greater risk of being seen than most other North American animals. In this connection, many people reason that a creature the size of bigfoot would be much more noticeable (i.e., there would be more sightings) if there were such a creature. This assertion is only partially correct. We believe that bigfoot is a primate. It therefore has the ability to crouch very low, fold-in his arms and reduce its visible size to one-half or less. A 6 foot (1.8m), 200 pound (90.6kg) man can reduce his visible size to 27 x 18 x 21 inches (68.6 x 45.7 x 53.3cm). If we use the same ratio for height reduction, this indicates that a 8-foot (2.43-m) bigfoot could reduce its physical height to about 36 inches (91.4cm). The other dimensions would not exceed the height, so for arguments sake we will assume they are the same. We therefore have a physical "presence" of something that is 36 inches (91.4cm) square. A shipping container of this size would fit easily into a modern recreational vehicle.

Many people who have had bigfoot sightings believe they at first saw a bear. When the creature stood up, they realized it was not a bear. The fact that the creature was "crouched" or bending over brought about the first impression. Considering the creature's ability to reduce its visible size, we can reason that there are probably many more bigfoot around than the sightings lead us to believe. This fact is applicable to both the presence of the creature in rural areas as well as "in the wild."

It appears highly reasonable that bigfoot know of human presence well in advance of actually seeing a person. The creature (like all creatures) would make itself as "scarce" as possible when threatened with confrontation. This factor indicates that trying to find a bigfoot creature in the wild may be bordering on impossible, unless the hunter "lucks out," as in the case of the sighting and filming of a bigfoot by Roger Patterson and Bob Gimlin at Bluff Creek, California in 1967. However, analysis of this incident indicates that other factors may have contributed to the sighting (noise from the creek, wind direction, men on horses rather than on foot). It might also be reasoned that the creature was totally preoccupied. Certainly all animals "drop their guard" now and then.

Chapter 6
Ohio Black Bears and Bigfoot

Few bigfoot investigators would argue that many bigfoot sightings in Ohio are, in reality, sightings of black bears. The Ohio Department of Natural Resources states that many black bears from other areas have taken-up residency in Ohio. They have been spotted in fourteen counties in the eastern part of the state. Wildlife authorities report that they see a consistent increase in the black bear population in Ohio. Many bigfoot sightings are made in areas that are known to have a considerable black bear population. It appears evident that people who are not familiar with these creatures and their habits, may believe they have seen something unusual.

The average American black bear *(Ursus Americanus)* is about 6 feet (1.8m) long. When it stands on all four legs it is about 2 to 3 feet (61 to 91.4cm) in height. Weight ranges between 225 to 500 pounds (101.9 to 226.5kg). They are solitary creatures, and eat just about anything, but are mainly vegetarians. They are not normally aggressive, but can be so if provoked. They are not normally seen upright, walking on their hind legs. However, they do walk upright in certain circumstances. When they do so, they are at least 6 feet (1.8m) tall and may appear even taller.

Bear footprints are somewhat "human" looking, but the big toe is on the outside, as opposed to the inside as with human beings and bigfoot creatures. Also, bear prints are generally much shorter than bigfoot prints. However, sometimes prints made by the front feet of a black bear are over-lapped with those of its back feet. This action results in a single impression that is larger than a print made by either of its feet (back or front) alone. Such tracks are termed "double tracked bear prints." A cast of such prints is shown here compared with a bigfoot print cast taken by Bob Titmus in California (1958).

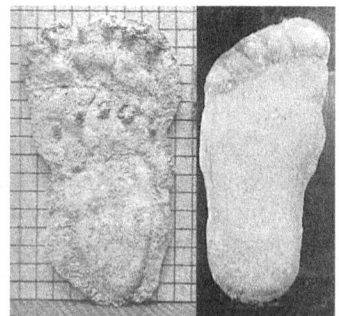

Cast of a double-tracked bear print (left) and a bigfoot print.

Confusion arises when, in the course of time, the entire double-tracked impression fades (back feet toes disappear), leaving what appears to be one large print. Many supposed bigfoot tracks have turned out to be double tracked bear prints.

In the photograph detail seen below, the *right* foot of the alleged bigfoot creature seen in the Patterson and Gimlin film (frame 61), is compared with a bear's *left* back foot (inset). We can see very clearly that footprints made by each creature would be distinctly different. Nevertheless, there is enough similarity to cause confusion, especially when footprints become "weathered." It should be noted that if the bear's *right* foot was shown in the illustration, the big toe would be on the opposite side.

Factors such as poor visibility, emotion, fright, excitement, over-reaction and so on are certainly conditions that would confuse the sighting of a black bear with the sighting of a bigfoot. However, having said that, there is a whole host of other information on sightings that does not "add up" to a simple encounter with a black bear. If a person states that the creature walked or ran a considerable distance on its "back legs," and looked like a gorilla, these factors point to a creature that is not a black bear. Further, sightings are often made by people who claim they are fully familiar with black bears and would know one if they saw one, regardless of the sighting situation.

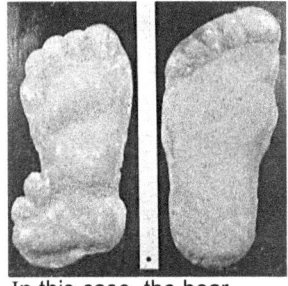

In this case, the bear prints (left) did not register, but we can see better similarity to the bigfoot print cast on the right.

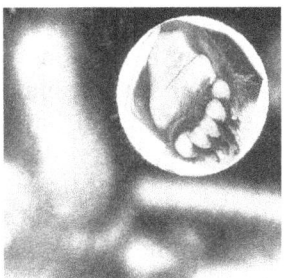

Comparison of actual alleged bigfoot's foot (left) and a bear's foot.

I believe the confusion of bears with bigfoot is much more prevalent with people in or from urban communities, as opposed to people in or from rural communities and wilderness areas. People in the latter are much more familiar with wild animals, and it is bordering on insult to tell these people they saw a bear when they inform they saw something totally different from a bear. People in authority, police and "experts," as it were, are often inclined to write off bigfoot sightings as bears because this is the most convenient explanation, and it discourages attention to the incident.

Chapter 7

Ohio Deer Kills and Bigfoot

There appears to be some evidence that bigfoot creatures hunt and kill deer as part of their food source. In this connection, Ohio, with its very large deer population, has emerged as a primary area for possible bigfoot related deer kills. Extensive research has been performed on this subject by Matt Moneymaker, a bigfoot researcher who resided in Akron, Ohio. The contents of this chapter are largely the result of his investigations. Your authors have not confirmed or verified the information provided by Moneymaker. The information, as follows, was gathered from Moneymaker's original website and is used here with permission.

Matt Moneymaker

In November 1992, Matt investigated a series of bigfoot reports that originated near the borders of Stark and Carroll Counties, Ohio. The incidents took place near a group of farmhouses that were situated on a large tract of land with 50-year-old second growth forest. The land had been reclaimed from earlier strip-mining activity.

The first sighting investigated involved a bow hunter who had shadowed through the woods a herd of about 15 deer. Just after sundown, the hunter waited at the edge of a field where deer where known to gather at nightfall. The deer grouped in the field as expected, and the hunter waited patiently for a buck to wander within the range of his bow. To his surprise, the entire herd was suddenly frightened and fled into the trees bordering the far side of the field. The hunter turned to his right and saw the reason for the alarm. A short distance from his position, he observed an upright walking creature in the shadows along the tree line. The creature, it appears, had also been stalking the deer, staying concealed in the trees.

The creature proceeded toward the hunter, and when about 35 feet (10.7m) away, noticed him. At this point it let out a high-pitched screaming wail or cry. The hunter stated that he felt the cry was far louder than one that could be made by a human being. It screamed four or five times, and then retreated inside the tree line. It was then heard to make a stomping sound before it proceeded deeper into the forest.

The hunter quickly retreated to a nearby house owned by one of his friends. He informed his friend of his experience, and a few days later this resident had a similar experience. While sitting in his kitchen he observed the deer again assembling in the field. As the deer quietly grazed, he heard a knocking sound coming from a wooded hillside overlooking the field. The sound was like a thick branch being hit against a tree trunk. Moments later, the sound appeared to be answered by a similar sound from a wooded slope closer to the house. He then heard a short, loud growling roar. With that, the deer bolted in panic into the trees at the base of the slope from where the answering knocks originated.

The following day, the resident went into the woods with his dog. He hiked up the nearby slope to a graded dirt road that led to a water tank. He observed two human-like footprints, 14 inches (35.6cm) long and 6 inches (25.2cm) wide, in the mud on the edge of the road. He continued up the road, and when he reached the water tank, his dog wandered into the trees. The resident followed his dog and it led him to the carcass of a young deer that had been recently killed. The resident further investigated the area and found two more young deer carcasses. Upon examining the finds, he observed that each of the deer had at least one broken leg. These breaks were very noticeable as the limbs were violently twisted and contorted. There were no bullet or bolt wounds, and the animals were basically intact except for the belly area that had been ripped open. The intestines were still attached, but had been pulled out of the belly and left in a pile beside the animal. The resident left the site, and upon arriving home, told his wife and neighbors of his findings.

Over the next few days he and others heard the unusual knocking coming from the woods at various times during the day and night. About one week later, the deer moved out of the area and the knocking discontinued.

Matt Moneymaker personally inspected the carcasses later with the resident, and asked him together with others in the area, if they knew of any animal that might kill and leave deer in the fashion described. They considered poachers, dogs, bears and cougars. None of these suggestions, however, could explain why the deer had just been left to rot. Poachers and animals kill deer for a reason, be it trophies or food. While a part of the deer might be left, certainly not the entire animal. A possible answer to this riddle became apparent with Matt's next investigation.

In December 1992 Matt responded to a bigfoot sighting in Guernsey County, which is about 60 miles (96.5km) south of Stark County. A Mennonite farmer in Guernsey reported seeing and hearing a bigfoot near his farm over the course of a few weeks. During the interview, the farmer asked, "Do those things *(bigfoot)* kill deer?" Somewhat surprised with this question, Matt responded, "Why do you ask?" The farmer then stated that he kept finding young dead deer down by a local creek. He could not find any evidence that the deer had been shot and was puzzled by the fact that the deers' legs had been broken and twisted around. His family had lived in that county for generations and did not know of any animal that would kill deer in this way, and then apparently leave them to rot. Matt expressed his interest in this occurrence and related to the farmer the incident in Stark County. Matt asked permission to return in a few days with his wife, a medical doctor, to inspect the deer carcasses. *(It might be noted that Matt's wife is also the daughter of two veterinarians and the granddaughter of a zoo curator.)*

Upon returning to the farm, Matt and his wife followed the farmer's directions and found one deer carcass. The carcass was in essentially the same condition as those found in Stark County – no evidence of bullet wounds, uneaten, intestines hanging out. Again, the wound that opened the deer's belly was not a clean cut. It appeared ripped, as if opened with a dull instrument or object. Dr. Moneymaker carefully inspected the open cavity and reported that the deer's liver was missing. This reddish-brown organ is normally located, and is highly evident, between the animal's rib cage and intestines. Dr. Moneymaker speculated that the predator merely open the deer's belly, moved away the intestines, lift up the rib cage, and remove the liver in its entirety. The Moneymakers then made a thorough search of the entire area. They found two other deer carcasses in exactly the same condition, and again with their livers missing.

ANALYSIS: When we analyze Matt's findings to this point, we see that some logical conclusions can be drawn from his information. The bow hunter's encounter appears to indicate that he happened upon a possible bigfoot who was scouting *(as opposed to hunting)* deer. The screams or cries made by the creature upon seeing the hunter were probably to scare away the competition. The knocking or signaling, as it were, may have indicated that two creatures were in place for a subsequent "scare and catch" tactic. Knocking a tree rather than vocal contact was employed so as not to prematurely frighten the deer. We can reason to some degree that the knocking required creatures with a hand to hold a piece of wood, which would indicate a primate of some sort. As for the general "scare and catch" tactic itself, it does not indicate any high intelligence. It is used by other animal predators.

As all of the deer killed had one or more of their legs broken, this indicates a possible method used to immobilize the animal. A bigfoot creature using a heavy tree branch might be able to accomplish leg breaks in a "rush" of deer. With a broken leg, a deer would be a fairly easy catch for a bigfoot. *Pure speculation we realize, but not beyond the realm of possibility.*

We next come to the issue of the abandoned deer carcasses and the missing livers. Why would a bigfoot, or any animal for that matter, take only a small part of its kill and leave the rest to rot? There was no evidence that the kill had been hidden for subsequent retrieval, so we cannot say the deer had been "stashed." Do any other predators do anything like this? Surprisingly, there is at least one – the grizzly bear. Towards the end of the summer when grizzlies have had their fill of salmon, they continue to catch the fish, but only eat its head. The rest of the fish is left on the bank or to float down the river where it is quickly snatched-up by gulls and ravens. The birds are well aware of the grizzlies' unusual habit, and patiently wait for their free meals. It is reasoned that the head is the most nutritious part of the fish, so the bear likes to "top off" his reserves with only this portion.

Can we possibly draw a parallel between grizzly bears and bigfoot, substituting deer livers for the latter in food selection? Matt Moneymaker did some fairly extensive research on deer liver to determine its nutritional value. His conclusions were as follows:

1. A fresh raw liver of a naturally fed young deer contains substantial amounts of every vitamin necessary for life, particularly those that would become naturally depleted in the fall and winter.

2. A deer's liver contains very substantial amounts of vitamin A, which is critical for an animal's ability to see in the dark. It is equally critical for the thickening of skin – callous formation effected through a process called keratinization.

3. A full 90% of all cholesterol produced in an animal is produced in the liver. Cholesterol is the building block for hormones.

4. Of all the organs, the liver secretes the highest amount of proteins into the blood. When this great repository of proteins is eaten and digested, the proteins are broken down into amino acids which are then used as the raw material to build new proteins and enzymes.

5. The liver is the primary repository for glycogen. Glycogen molecules are basically ready-to-use energy storage molecules. In the digestion process, these glycogen molecules get broken down and absorbed as glucose that can be used by the muscles and other tissues, such as brain cells, or can be rebuilt into ready-to-use energy storage molecules in the predator's own liver.

6. The liver stores tremendous amounts of lipids, another excellent calorie source.

With regard to the physical deer liver itself, it is noted that this organ is relatively easy to remove from a carcass. It is also easily eaten and digested. Its average size, would be about the same as a 5-pin bowling ball. For a large primate, the organ would be a natural, and perhaps exclusive choice, as an essential and convenient food source. In terms of maximum nutritional gain for energy expended, concentrating on deer livers, and ignoring the rest of the animal, would make very good sense for a bigfoot creature.

There is one other possible connection. It may be that the deer livers were being obtained for medicinal purposes. Various animals are known to seek certain plants strictly for their medicinal value as opposed to their food value

(Chimpanzees are particularly noted for this practice). In a bigfoot sighting report from Alberta, British Columbia, it was stated that the creature was eating devil's-club *(Oplopanax horridum)*. This large, prickly shrub is of the ginseng family and is found in the west from Alaska to California. It is used in herbal medicine (your author recalls its use many years ago in connection with sugar diabetes). Ginseng itself, of course, is well-known for its purported medicinal value. It is not inconceivable that a higher primate, such as bigfoot, may have graduated to include animal organs for the treatment of various disorders. In this case, we can easily justify the creature removing just a deer's liver and disregarding the rest of the animal. Let's face it, human beings do the same thing (bear's feet in British Columbia, rhinoceros horns in Africa).

Matt's next deer kill investigation took place in the summer of 1993. A retired machinist who lives near Wills Creek, Coshocton County, related his encounter with a Bigfoot in the fall of a previous year. He was so taken by his experience that he built a statue of the creature, and placed it in front of his cabin with a sign reading, "SASQUATCH VALLEY." He told Matt that while out hunting, he hiked trails into some of the more remote, uninhabited hollows of the Wills Creek area. He found some deer tracks and followed them, eventually coming to a cliff with an overhang forming a cave. He entered the cave and found a number of severed deer legs. The legs were neatly arranged side-by-side. No other animal parts were present. He decided to rest there for awhile, and a short time later a bigfoot entered the cave from one side of the overhang. Paralyzed with terror, the old machinist could not believe his eyes. He just sat there and stared at the hairy creature, who stared back at him from a few yards away at the lip of the overhang. The machinist eventually stood up and slowly made his way out of the cave on the opposite side of where the creature was standing. The bigfoot just stood there, and did not pursue the intruder as he quickly retreated into the woods.

ANALYSIS: From this account, we can gather a few more pertinent facts. It appears that deer are a food source for bigfoot creatures, and that they do stash their food. As only the deer legs were evident in the cave, it may be reasoned that the remainder of the edible parts were eaten (or partially eaten) at the kill site (less desirable parts simply left). The legs, being the easiest part to transport, were taken for future consumption. The fact that the bigfoot was occupying a cave is also noteworthy. While we *suspect* that the creature is a cave-dweller, we really don't have a lot of information to substantiate this thought. The finding of possible bigfoot beds and nests point to an different life-style, however, a combination of several styles is not inconceivable.

In the fall of 1993, Matt followed up on a number of bigfoot sightings near Berlin Lake, Portage County. He brought up the subject of unusual deer kills with a county sheriff's deputy. The deputy stated that a few years previous, wildlife officers had found several deer carcasses close together near the edge of Berlin Lake. The animals had been mutilated; however, the deputy did not have any specifics in this regard. It was learned that poachers had been ruled out as a cause, which indicates there were no bullet wounds in the animals. The only plausible explanation given was activity by devil worshippers.

Later that year, Matt was contacted by an acquaintance in the Berlin Lake area who had found the carcass of a young deer a mile or so north of the lake. Matt inspected the carcass, and observed that it was in the same state as the others he had seen (intestines removed and piled along side the body); otherwise, the animal was intact. He was unable to inspect the cavity for presence (or non-presence) of the liver because the carcass was frozen solid. He later spoke with a resident in the area, who lived about 100 yards (91m) from the kill. Matt asked the resident if he had seen any cougars or bears in the area of late, or if he had seen anything unusual. The resident stated he had not seen any of these animals in the area. However, the week before, his wife had seen large human footprints in the snow close to their barn. The couple were at a loss to explain the prints, as it appeared the trespasser was walking barefoot in snow with the temperature below zero.

ANALYSIS: From the second Berlin Lake incident, we can gather that bigfoot creatures probably continue hunting throughout the winter months. There is no indication that the deer's legs were broken in either Berlin Lake incidents, so the kill method did not appear to include this action.

The next major Ohio deer kill incident took place in Columbiana County in October 1994. Matt received a telephone call from a resident informing him that over the course of two weeks, people had found four young deer carcasses in the woods near their homes. Matt went on location again and observed that the carcasses were in the exact same condition as those in the Stark and Carroll Counties area. This time, he video-taped the animals, drawing attention to the broken legs and missing livers.

Later that year, Matt heard again from his contact in the Berlin Lake area. Relatives of this resident were visiting from Wisconsin and had a sighting on their return home. As they were driving through central Wisconsin at night, they nearly hit a bigfoot crossing the road. They observed that the creature had the body of a young deer slung over its shoulder.

Over the years, Matt has broached the subject of deer kills with other bigfoot researchers. Joan Jeffers, a researcher in Pennsylvania, has a report of a group of people seeing a bigfoot carrying a deer on its shoulder. Another Pennsylvania report involves a hunter who shot a deer and readied it for gutting by hanging it on a tree. He went to his car for supplies, and when he returned the deer was gone. Leading away from the scene was a trail of deer blood and large human-like footprints.

There have also been deer kill finds in Illinois. Here, several mutilated deer carcasses were found stacked in a forest. Further, there is the testimony of an elderly lady who grew up in the upper Coppei region of the Blue Mountains, Washington state. The lady states that bigfoot reputedly chased elk over the brink of Coppei Falls. The bigfoot would then retrieve the dead or injured animals.

> **Columbiana Sightings**
>
> Matt had been to this location in the summer of 1994 to investigate sightings, and hopefully record vocalizations that were being attributed to bigfoot. Residents in the area informed Matt of sightings extending back to the mid-1980s. Nothing, however, occurred during Matt's visit.

The question as to how a creature the size of bigfoot could sustain itself in North America is often asked. As we do not believe the creature hibernates, it must obtain food all year round. Sustenance during winter months would be difficult for bigfoot just as it is with all other wild animals. The other seasons would provide fish, adequate vegetation and wild fruit, so would not present a problem.

Fish would probably form a main part of a bigfoot's diet in these seasons as it does with bears. Many sightings involve the creature in a shallow river, apparently looking for fish. Remarkably, in Bella Coola, British Columbia, First Nations people claim bigfoot creatures throw rocks at fishermen who are net fishing in rivers. This action gets the attention of the fishermen who will throw a fish up on the shore to appease the creatures. The natives state that the fish are quickly taken. Many First Nations people in this area are firm believers in bigfoot and have no doubt that the creature exists.

In that deer are a year-round food source, it is reasonable to assume they form a critical part of a bigfoot diet in the winter. From the information we have presented, it appears bigfoot are very capable deer hunters. Areas with large deer populations, therefore, may be the most likely for people to observe bigfoot creatures.

Further, it is interesting to note that a possible bigfoot method or means to ensure food availability has been formulated by two noted British Columbia Provincial Museum (Royal Museum) authorities. In his book, *BIGFOOT – Startling Evidence of Another Form of Life on Earth Now* (Berkley Publishing, 1974), Dr. John Napier states the following (page 172):

> Frank L. Beebe and Don Abbott of the Provincial museum, Victoria, B. C., have come up with a most ingenious "model" for the Sasquatch's feeding habits. They base it on the life-style of the wolverine, a weasel-like mustelid of large proportions (3ft. or more including tail). This extraordinarily interesting animal has a wide home-range of 200 miles or more, is largely carnivorous and rapacious with it (not for nothing is it known as the 'glutton'). Wolverines, broadly speaking, occupy the same habitat as the Sasquatch. A particularly relevant aspect of their behaviour is that wolverines cache their food in natural 'deep-freeze' lockers above the snow line, winter and summer. They range upwards into the snowfields when food is scarce at lower altitudes, open their lockers and (presumably) carry the food down to altitudes where it gradually thaws. This model might explain how the Sasquatch survives through the winter, and why so-called Sasquatch footprints have been observed at high altitudes by skiers and snow-mobilers; equally of course, the habits of the wolverine might account for the very existence of these amorphous tracks. Let me give full rein to imaginative speculation: could deep-freeze behaviour patterns also explain the apparently inexplicable occurrence of Yeti footprints high above the snowline in the Himalayas?

Chapter 8

The Patterson Legacy

Certainly, the single greatest and most convincing evidence of the existence of Bigfoot is the film taken by Roger Patterson and Bob Gimlin at Bluff Creek, California in October, 1967. Unlike the evidence for some of the other "great mysteries" in the world, this film is generally withstanding the test of time. In short, there is still as much evidence to support its authenticity as there is to discredit it. There have been no death-bed confessions, moldy monkey suits or secret diaries that have surfaced over the 37 years since the film was taken. Roger Patterson died in 1972, never deviating in the story of his remarkable experience. Bob Gimlin lives in Yakima County, Washington State, still firm and steadfast that the two adventurers saw and filmed a genuine bigfoot.

To be sure, major factors supporting the film are the numerous bigfoot sightings and footprints. What may be considered "reliable" reports and findings pre-date the film by over one hundred years. It was, in fact, sighting reports and footprints that influenced Patterson to take up his search for the elusive creature.

Patterson's first exposure to the bigfoot phenomenon was an article he read by Ivan T. Sanderson that appeared in *True* magazine, December 1959. The article related the finding of large humanoid footprints by Jerry Crew in the Bluff Creek, California area the previous year. Sanderson, who had long been involved in the search for humanoid creatures, gave considerable credibility to Crew's finding. In particular, he called attention to the vast unexplored regions in the northwestern part of California. His closing paragraph in the article is noteworthy:

> Before you are tempted to scoff and put this story down, bear in mind a few things. This area extends over 17,000 square miles, and nobody lives there.

Apart from the higher ridges and mountain peaks, the ground area is completely concealed from the air by forest. It has never been properly surveyed or mapped. Yet for all of this, the area is well watered, overgrown with berries, full of small game, and never completely snowed in. Though it nestles in the midst of civilization, and is fertile and livable as [that of] civilization, it is completely uncivilized. Almost anything could be living there. From the evidence, something is. Will somebody please do something about it before it is too late?

Patterson started his research by talking with First Nations people on the Yakima Reservation (Washington State). Native belief in bigfoot (for which Natives have numerous names), gave him further encouragement. Sometime in 1964 he ventured into the Bluff Creek, California area to look for the creature. Acting on information from Pat Graves, a Forest Service worker in the region, Patterson saw first-hand recent 17- inch (43-cm) tracks in the Laird Meadow Road vicinity, similar to those reported by Jerry Crew. Patterson made a cast of one of the prints as seen here. He was greatly moved by this experience, as can be seen by his own words, "I was so astonished I could only stare and try to picture the creature that had made those tracks only the day before."

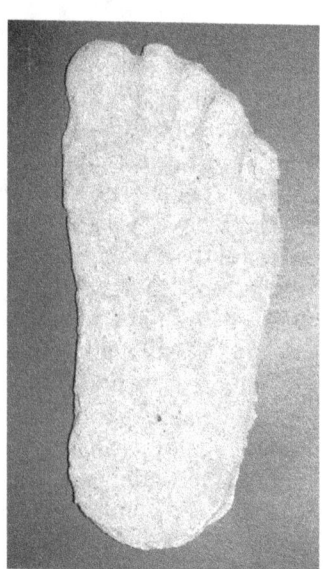

Copy of a cast made by Patterson of a footprint found near Laird Meadow Road.

This experience infused in Patterson a burning desire for more knowledge. He continued his research, and collected numerous newspaper and magazine articles on the subject. An artist of considerable ability, he drew pictures of the events he read about, or which were related to him by others. All of the information he gathered, together with his own thoughts and conclusions, he compiled into a book entitled, *Do Abominable Snowmen of America Really Exist?*

(Franklin Press Inc.) that was published in 1966.

Patterson's incredible experience the following year (October 20, 1967) in which he and Bob Gimlin personally saw and filmed an alleged bigfoot is summed up by Patterson as follows:

Last Friday, my companion – Bob Gimlin, a part Apache fellow who's good at tracking and so on – and I started up an old logging road where a particular lot of big tracks had been seen. Some of the tracks were 17 inches long. We rode horses, and I had a 16 millimeter movie camera in my saddlebag. We both had high-powered rifles but we agreed that if we found a sasquatch we wouldn't shoot unless we absolutely had to. About 1:30 in the afternoon, as we rounded a bend in the road, we saw the creature. My horse reared, and then fell as I tried to control it. But I got the camera out and yelled to Bob to cover me with his rifle while I tried for the pictures. The thing was across the creek beside the road, about 50 yards away. I ran down to the creek and got on a high sandbar to film it. It was obviously a female, for although it was covered in hair you could see it had large breasts. It stood about six feet tall, maybe more, and was very broad. We figured the weight at somewhere between 350 and 400

Above, a model of the film site (reasonably to scale) as it appeared on October 20, 1967. Roger Patterson mainly filmed the creature from the position indicated with a map pin. The model depicts the moment when the creature turned and looked at Patterson and Gimlin, as seen in the lower photograph (frame 352 of the film).

pounds. She stood there for maybe half a minute, and then started walking away, still upright. She crossed the creek, got back on the logging road up ahead and moved out of sight. Bob started to follow on his horse, but I called him back. The tracks we'd seen earlier indicated she was part of a family group, and that could be dangerous. I was shaking quite a bit, so the film isn't too steady, but it shows the thing clearly. I've believed they existed for a long time, just to talking to many eyewitnesses. Now there's no doubt at all.

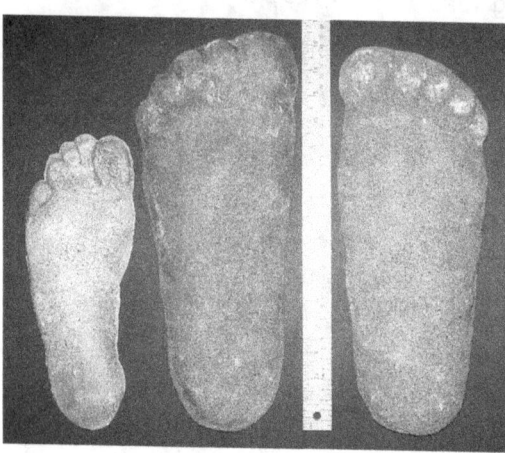

Casts of footprints made by the creature filmed by Roger Patterson and Bob Gimlin. A cast of a human foot (11.5 inches (29.2cm) long is shown on the left for comparison purposes. The creature's prints in the soil were about 14.5 inches (36.8cm) long. (Bigfoot casts shown are direct copies of those taken by Patterson.)

Roger Patterson's untimely death, just one month short of his 40th year, brought to a close his dream of actually capturing the creature he had filmed. Considerable research, however, has been performed on the film by many professionals and dedicated bigfoot researchers.

In 1998 the North American Science Institute, under Jeff Glickman, a forensic examiner, completed a highly intensive analysis of the film. The following is a summary of the findings.

1. Measurements of the creature:
   Height – 7 feet, 3.5 inches, (2.2m);
   Waist – 81.3 inches (206.5cm);
   Chest – 83 inches (210.8cm);
   Weight – 1,957 pounds (886.5kg);
   Length of arms – 43 inches (109.2cm);
   Length of legs – 40 inches (101.6cm).

2. The length of the creature's arms is virtually beyond human standards, possibly occurring in one out of 52.5 million people.

3. The length of the creature's legs is unusual by human standards, possibly occurring in one out of 1,000 people.

4. Nothing was found indicating the creature was a man in a costume (i.e., no seam or interfaces).

5. Hand movement indicates flexible hands. This condition implies that the arm would have to support flexion in the hands. An artificial arm with hand movement ability was probably beyond the technology available in 1967.

6. The Russian finding on the similarity between the foot casts and the creature's foot was confirmed.

7. Preliminary findings indicate that the forward motion part of the creature's walking pattern could not be duplicated by a human being.

8. Rippling of the creature's flesh or fat on its right side was observed indicating that a costume is highly improbable.

9. The creature's feet undergo flexion like a real foot. This finding eliminates the possibility of fabricated solid foot apparatus. It also implies that the leg would have to support flexion in the foot. An artificial leg with foot movement ability was probably beyond the technology available in 1967.

10. The appearance and sophistication of the creature's musculature are beyond costumes used in the entertainment industry.

11. Non-uniformity in hair texture, length, and coloration is inconsistent with sophisticated costumes used in the entertainment industry.

> NOTE: There is much controversy over the height and weight of the creature. Nevertheless, it appears that it was at least 6 feet, 6 inches (1.98m) tall, and weighed at least 542 pounds (245.5kg).

In other areas of bigfoot research, footprints indicate a very straight-line footprint pattern. Human tracks, in comparison, have greater alternation between footprints. This difference may have something to do with the weight and structure of the creature. It has now been reasonably confirmed that the creature in the Patterson/Gimlin film also had a straight-line footprint pattern. Noteworthy in this regard, I have been given to understand that people in certain North American First Nations tribes walk in this fashion.

Whether one believes in the film or not, there is no doubt that the Patterson legacy brought tremendous, much needed focus on the bigfoot issue.

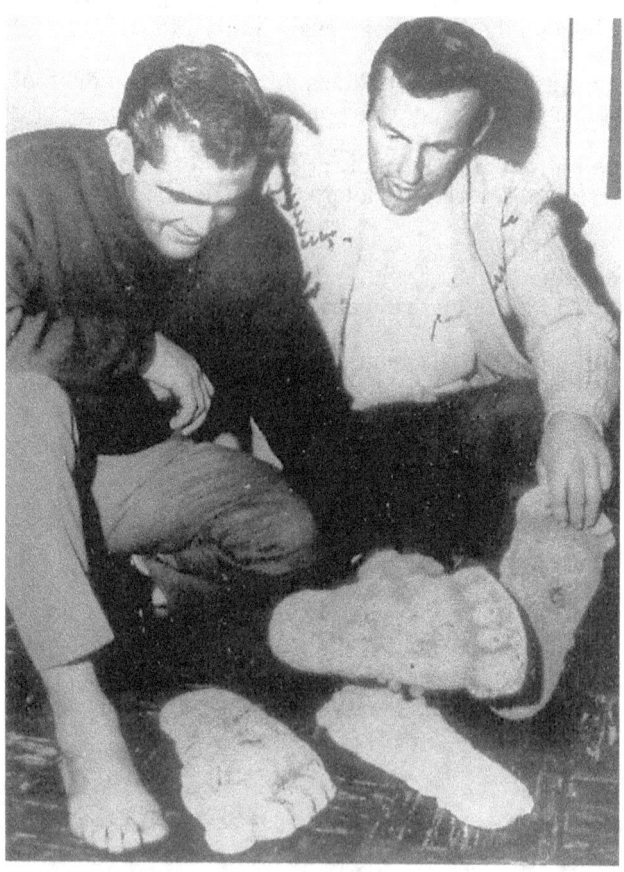

Bob Gimlin (left) and Roger Patterson (right). Gimlin is comparing his foot to the Laird Meadow Road area cast (1964). He and Patterson are holding casts made from prints left by the creature at the Bluff Creek film site.

Chapter 9

Intriguing Questions

There are certainly some intriguing questions on possible bigfoot existence. The following are a few of the most pertinent issues, together with my thoughts.

Where are the bones? This is about the most common question asked of bigfoot. What is being asked here is, when one of the creature dies (old age, accident, disease) should we not be able to find its bones? Remarkably, the same question can be asked of any other large animal, even birds. We know, for example that there are quite a few cougars in the Pacific Northwest. However, if one were forced to prove this creature's existence by *finding in the wild* and presenting its skeleton (or part thereof), it would be very difficult to do so. I will mention here, that while I have used the cougar as a example, the odds of finding the skeleton of a wolverine, a far less plentiful creature, would be even more staggering.

The reasons for the scarcity of such skeletons (or any skeletons) is that animals about to die conceal themselves, and then when they do die, their remains are quickly consumed by other animals. Certainly, bones of prehistoric creatures have been found, but here the circumstances were such that the bones were preserved. In other words, the creatures fell into a tar pit or a bog, or became immediately buried (landslide or avalanche) – all of which acted to preserve their bones. Bigfoot bones would also get preserved in this way, but none of the creatures appears to have died under such circumstances (its intelligence might play a part in this regard).

If bigfoot is part human, why has it chosen to remain primitive? The intended argument here is that humans, by their nature, evolve to make things easier on themselves (invent things, utilize tools and so forth). This being the case, why has not bigfoot done the same?"

As the old saying goes, "necessity is the mother of invention." When human beings are content with the situation (be that what it may) they do not put forth effort to change it. Many people on the earth today are living in the same conditions as they did centuries ago. Indeed, had Europeans not settled in North America, the First Nations people who were here would probably still have the same living conditions today (they would not have invented refrigerators and washing machines so to speak). It is evident that bigfoot is content with its existing way of life.

How do we rationalize the fact that bigfoot does not use fire? First of all, if bigfoot did use fire, it would have been captured, and probably exterminated, hundreds of years ago. While there might be an argument on the lack fire usage if the creature is part human, it is weak. Human beings have not always used fire. The probably answer to this question is that the creature does not _need_ to use fire. In other words, it does not need fire for warmth and does not need (want) to cook its food. Moreover, it must be kept in mind that human beings are the only animals that use fire. This fact might be a clue that bigfoot is just another animal, no more closely related to human beings than other known primates.

Primates are social creatures. How do we justify so many sightings of a single bigfoot? My first thought here is that the creature is not alone – another or others are nearby. My second thought is that some of the creatures we are seeing are indeed "rouge" creatures that have separated themselves from a group. Finally we can reason that perhaps the species is not highly social. Nevertheless, it needs to be mentioned here for clarification that there have been sightings of more than one bigfoot, and also family groups (i.e., mother, father, and child or children).

Chapter 10

Where to From Here?

The evidence we have for the existence of bigfoot creatures in North America is impressive. We have *alleged*:

Sightings
Motion Picture Film, Videos and Photographs
Footprints
Handprints
Body Impressions
Hair
Feces
Beds/Structures

The most recent hard evidence is a body impression cast taken in Washington in September 2000. Researchers with the Bigfoot Field Researchers Organization set out fruit in an area of soft earth/mud in the Skookum Meadows area (Gilford Pinchot National Forest), and it appears a bigfoot semi-reclined on the ground and reached for the fruit. A cast was made of the impressions, now known as the Skookum cast. In one spot it appears the creature dug in its heel. The photograph seen here shows a cast of the heel and the back of the creature's leg. This cast was made from the heel impression in the body cast. The heel portion is about 4 inches (10.16cm) wide. Professionals examined the body cast and concluded that it appears to be from impressions made by an unclassified primate.

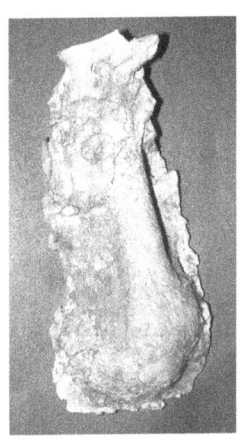

Cast of a possible bigtoot heel impression.

Despite all this evidence, the creature's possible existence is not recognized by the general scientific community. As a result, a proper government funded investigation has never been undertaken to find a bigfoot. Certainly, as Dr. Grover Krantz pointed out, if such evidence were presented on the existence of some other non-classified animal (other than a primate), scientific recognition would be very high and government funded research would quickly commence.

The connection of evidence with a North American non-human primate, however, is not the only issue for scientists. Over the years bigfoot has become a "fun thing," with numerous hoaxes perpetrated by people craving attention, or trying to outsmart the dedicated researchers and few professionals in the field. I know of two major hoax initiatives that fully succeeded in this regard.

We also have the convergence of what might be termed the "new sciences" that uses non-conventional logic to put forth theories on unexplained phenomena. Any self-respecting scientist who jumps into this quagmire is at risk of committing professional suicide. It is really no wonder that most scientists demand a bigfoot body, part of same or bones before they will get involved. For such evidence, however, we really need a proper investigation, so around and around we go.

The only real hope lies in a chance discovery of bigfoot remains by an amateur researcher, a hunter mistaking the creature for a bear, or a road kill. Nevertheless, I do believe that the evidence we have coupled with more/better photographs of the creature would be enough to tip the scale for reasonable scientific involvement.

Chapter 11

Mysterious Tracings

People with animal-like characteristics were eagerly sought during the 19th century and used as side-show attractions. While this practice was deplorable, it nevertheless provided us with a record of such people. The fact that they existed (and are still found) fuels speculation of evolutionary "throw-backs." Is it possible these people somehow reflect human origins? If this thought is considered feasible, could it be possible that some of man's distant relatives still inhabit the earth on a different evolutionary branch?

Moreover, reports of hominids beyond North America add to the mystery and raise speculation that perhaps bigfoot and these other creatures are somehow related.

KRAO: A young Siamese girl called Krao (c1876-1926), who was billed as *Darwin's Missing Link*, was exhibited by the Ringling Brothers in Europe during the early 1880s. The six-year-old girl is seen here with her circus manager. In addition to her ape-like physical features, two well-known physicians reported that she had extraordinary prehensile powers of the feet and lips. Although unquestionably human, Krao was certainly one of the most unusual people in recorded history. She eventually became a well-read lady who spoke several languages. She

Krao with her circus manager.

died at the age of forty-nine.

JULIA PASTRANA: A strange and sad connection may be made with Julia Pastrana (1842-1860), a highly popular side-show attraction. Her remarkable appearance made many people think she was not completely human. For the following account of Julia's life and subsequent history, I have drawn heavily from a book by Dr. Jan Bondeson entitled, *A Cabinet of Medical Curiosities* (Cornell University Press, New York, 1997).

Pastrana's early history is shrouded in mystery. We are told that Espinosa, an Indian woman (a so called Root-Digger Indian) in the Sierra Madre, Mexico, became separated from her tribe in 1830, and was believed to have drowned. However, in 1834 (or 1836) a group of cowboys found her in a cave together with a two year-old female child. Espinosa told the cowboys that she had been captured by hostile Indians and imprisoned in the cave. She stated that while she loved the child dearly, it was not her child. There were no other human beings in the area where she was found, which was said to be, "a region of country abounding in monkeys, baboons and bears."

Photograph taken in 1860 of Julia Pastrana's preserved corpse (mummy).

Back with her people, Espinosa continued to care for the child, who was christened Julia Pastrana. We next learn that after Espinosa died, Julia was taken in as a servant girl by the governor of Sinaloa. The governor wanted to study her as a curiosity. In 1854, Julia decided to return to her tribe after being mistreated. On her way back to her people, she was spotted by an American named M. Rates, who persuaded her to go to the United States and be exhibited for money in a sideshow. Rates billed her as the "Marvelous Hybrid or Bear Woman," and she became a major attraction.

In New York, Julia was examined by Dr. Alexander B. Mott, who declared her to be "one of the most extraordinary beings of the present day," a hybrid between human and orangutan.

Under a new manager, J.W. Beach, Julia was exhibited in Cleveland. Here, she was seen by Professor S. Brainerd, M.D., who concluded that her hair, skin and protruding jaws, "entitle her, I think, to the rank of *distinct species.*"

Now with a third manager, a man by the name of Lent, Julia toured throughout Europe. During this time, we are also told that another examination (possibly by Charles Darwin) revealed that her teeth were aligned like those of an ape, rather than the normal human alignment. Julie married Lent and she soon gave birth to a male baby while in Moscow, Russia, who resembled her in every way (1860). Both mother and child died shortly after the birth. The baby lived for 35 hours. Julia died about 3 days after her baby died. Lent had the bodies of both Julia and the baby mummified and exhibited them. The photograph seen here, taken in the early 1860s, is of Julia's mummy.

The two mummies were last exhibited in the early 1970s (including an American tour in 1972). Public outcry prevented further exhibits, whereupon the mummies were put in storage at a fairground near Oslo, Norway. In 1976 thieves broke into the storage building and extensively damaged Julia's mummy. The baby mummy was thrown into a ditch outside and totally destroyed by mice. In 1979, thieves again broke into the building and this time took Julia's mummy. Remarkably, police recovered the complete artifact a short time later after children reported that they had found a human arm in an Oslo suburb dump. The police did not notify the legal owner of the mummy. They turned it over to the Institute of Forensic Medicine, Rikshospitalet, Oslo, where it still resides at last word.

Dr. Jan Bondeson viewed the artifact in 1990 and provides the following information in his book.

> When I saw Julia Pastrana's mummy in 1990, it was standing on a small wooden board covered with fabric. The right arm had been torn off and lay in front of the

mummy; the right side of the face had been torn open as well, and the eye on that side was missing. The Russian dancer's costume placed on the corpse in 1860 had been torn off by the thieves, and the mummy was completely unclothed apart from the remains of the original boots. The hairy growth was greatly diminished by the ravages of time, but the abnormal hairiness of the forehead was still evident and parts of the whiskers were also preserved. The skin was dark brown and parchment-like.

It would be interesting to apply modern technology to Julia to determine the credibility of the claims of early scientists.

ZANA: The story of Zana, a Russian ape-woman, is truly remarkable. Zana died in the 1880s or 1890s, so some people in the area where she lived actually remembered her when questioned by researchers in 1962. It is believed hunters captured her in the wild whereupon she was sold. She changed hands several times, and eventually became the property of a nobleman. The following description of Zana is quoted from Dmitri Bayanov's book, *In the Footsteps of the Russian Snowman* (Crypto Logos Publishers, 1996).

Zana being taunted by children. (Illustration by Lydia Bourtseva)

> Her skin was black, or dark grey, and her whole body covered with reddish-black hair. The hair on her head was tousled and thick, hanging mane-like down her back. From remembered descriptions given to Mashkovtsev and Porshnev, her face was terrifying;

broad, with high cheekbones, flat nose, turned out nostrils, muzzle-like jaws, wide mouth with large teeth, low forehead, and eyes of a reddish tinge. But the most frightening feature was her expression, which was purely animal, not human. Sometimes, she would give a spontaneous laugh, baring those big white teeth of hers. The latter were so strong that she easily cracked the hardest walnuts.

Zana was trained to perform simple domestic chores, and became pregnant several times by various men. Remarkably, she gave birth to normal human babies, four of whom survived to adulthood (two males and two female). The youngest child, a male named Khwit, died in 1954. All of the children had descendants.

Zana cuddles her first newborn. Immediately after birth, she washed her infants in a cold spring. Unable to stand the shock, they died. Villagers thereupon took newborns away from her. We might reason Zana was acting on instinct, but did not comprehend that her infants were half-breeds. Had they been totally of her kind, they would have probably survived.
(Illustration by Brenden Bannon)

Several expedition were made in the 1960s and 1970s (notably those headed by Professor Boris Porshnev and later Igor Bourtsev) to find Zana's grave and exhume her remains for examination. While many sites were explored, the researchers were unable to find a skeleton that matched the description of Zana. On the Bourtsev expedition of 1978, it was decided to exhume the remains of Khwit, whose grave was well indicated. The idea, of course, being to determine what traits he had inherited from his mother. Igor is seen here at the grave site holding Khwit's skull. The skull was taken to Moscow for study by anthropologists.

While Russian anthropologists reported that the skull was different from that of ordinary human beings, such was not the opinion of Dr. Grover Krantz, an American anthropologist. Krantz stated that the skull is from a fairly normal, modern human.

Igor Bourtsev examines Khwit's skull at the grave site.

Khwit is seen in this photograph. He was extremely strong, difficult to deal with, and quick to pick a fight. He lost his right hand as a result of one of his many fights with fellow villagers.

THE KARAPETIAN HOMINID: We are told that in December 1941, a Russian army unit in the Caucasus observed a strange hairy man near their post. Fearing that he might be with the enemy, soldiers quickly captured him. Because of the man's unusual appearance, Lt. Col. V.S. Karapetian, a doctor in the Army Medical Corps was asked to examine him. The following is Dr. Karapetian's statement made to a magazine correspondent on the incident.

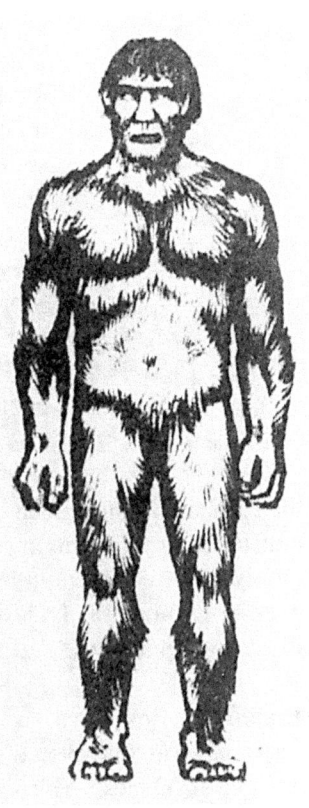

The man I saw is quite clear in my memory as if standing in front of me now. I was inspecting him on the request of local authorities. It was necessary to establish whether the strange man was an enemy saboteur in disguise. But it was a totally wild creature, almost fully covered with dark brown hair resembling a bear's fur, without a mustache or beard, with just slight hairiness on the face. The man was standing very upright, his arms hanging down. He was higher than medium, about 180 centimeters. He was standing like an athlete, his powerful chest put forward. His eyes had an empty, purely animal expression. He did not accept any food or drink. He said nothing and made only inarticulate sounds. I extended my hand to him and even said 'hello.' But he did not respond. After inspection I returned to my unit and never received any further information about the fate of the strange creature.

In providing more details at a later date, Karapetian revealed that the man was cold-resistant, and preferred cold conditions to

Lt. Col. V.S. Karapetian, MD.

normal room temperature. He was shown to Karapetian in a cold shed and when the doctor asked why he was kept in such cold conditions, soldiers informed that he had perspired excessively in the building where he was first taken. Elaborating on the man's face, Karapetian stated that the subject had a very non-human, animal-like expression. Moreover, Karapetian revealed that the man had lice of a much larger size and of a different kind than those found on humans. The doctor informed the authorities that the entity was not a man in disguise but a "very, very wild" subject with real hair. The drawing of the Russian hominid seen here is said to show the appearance of the unusual man. The drawing was not made Dr. Karapetian as is commonly believed. It could have been made under his direction; however, it is more likely just an artistic rendering made for general illustration purposes.

BASSOU: Bassou, a so called *ape-man,* was discovered in the Valley of Dades, Morocco in 1937. Bassou has massive bony ridges above his eyes and a sharply receding forehead. His lower jaw, teeth, chin and cheekbones are greatly pronounced and "ape-like" in appearance. His arms are so long that his fingers reach below his knees when he is standing upright. He is totally shunned by the people of valley for superstitious reasons related to his ape-like appearance. It is said that he sleeps in the trees and subsists on dates, berries and insects. He normally does not wear any

Bassou, c. 1974.

clothes or coverings, and is unable to speak any language. No one knows where he came from. Were he born in the mid-1800s, he would certainly have been eagerly sought as a sideshow attraction, and undoubtedly billed as another *missing link*. Bassou appears to be a true ape-man. Through some genetic miscalculation, he has possibly ended-up with some of the characteristics of our earliest ancestors. Bassou is different, but certainly not deformed. His unusual appearance is natural. He is fully capable of sustaining himself in complete harmony with nature. It is clear that Bassou is a very special person.

CHINESE/BIGFOOT CROSS-BREED: In October 1997, an interesting bigfoot related article appeared in the *World Journal* newspaper which is published in Taiwan.

The article states the following information (not a direct translation).

A woman, who works for the Bigfoot Research Center in China, was going through the belongings of her recently deceased father. Her father had been with the Wildlife Research Center in China. Among the belongings she found a video tape taken in 1986 that contained footage of an unusual person in a very remote forested area of China. The person, a male, about 33 years old, was very tall (about two meters *or 6 feet 5 inches*). He had a small head, and what appeared to be a kind of tail. His body shape and arms and legs were similar to those of the North American bigfoot. He did not have any noticeable long hair, and did not speak any language. He took fairly large steps when he walked. The mother of the "boy" was still alive when the video was taken. The mother stated that she had been kidnapped or abducted by a "wild man" after the death of her husband, and the boy was an offspring of her relationship with the wild man. The woman previously had a son by her husband. This son was an officer in the army, and he persuaded his mother to tell her story to the Wildlife Research people. She told her story under the condition that the research people would not reveal her identity while she was alive because she was ashamed of what had happened.

The article goes on to state that Chinese wild men have been recorded as far back at 100-200 B.C. It also mentions a monkey-boy who was discovered in 1932, but its existence was not reported until after it had died.

FRANCIS DE LOYS' MAN-BEAST: The unusual creature seen here has been making the rounds in the field of cryptozoology for 85 years. It is alleged the Swiss geographer Francis de Loys (or a member of his expedition) shot the creature in the jungles of Columbia, Venezuela in 1920. We are told that two of the creatures attacked the team, forcing the men to defend themselves. One of the creatures fell, and the other fled. De Loys' took this photograph and then removed the creature's skin and head, which he cleaned in preparation to transport them back to Switzerland for scientific study. On the

Francis de Loys' controversial photograph of a creature not known to science.

return journey, the artifacts were lost in a boating accident. All that remained for proof of the creature's existence was the photograph.

De Loys' claimed that the animal shown was just over 5 feet (1.5m) tall. This claim was disputed until it was noticed that the creature was sitting on a fuel crate of known dimensions. Using the height dimension of such a crate, it was possible to confirm de Loys' claim.

The creature's height and unusual human-like expression sets it apart from other primates known to science. The find was given added credibility by eye-witness accounts in 1931 of a similar creature in British Guiana (now Guyana). Then in 1968, the explorer Pinto Turolla reported that he saw a similar creature in the area of Marirupa Falls, eastern Venezuela. He also had a second brief sighting in 1970 on the eastern slopes of the Andes in Ecuador. Moreover, in 1987 mycologist Gary Samuels reported he also saw possibly the same creature in Guyana.

Most authorities contend the creature seen in de Loys' photograph is nothing more than a larger than average spider monkey. The evidence presented here, however, seems to support de Loys' claim.

THE YETI: Said to inhabit the Himalayas, the yeti was first brought to the attention of the outside world about 100 years ago. Since that time, many expeditions have been undertaken to find the creature but all have failed in this quest. There are many documented sightings, some very credible, but absolutely no photographic evidence. Alleged footprints in the snow remain the main tangible evidence of the creature's existence. The yeti footprint cast (copy) shown here was created from a photograph. The cast is about 12.5 inches (31.8cm) long.

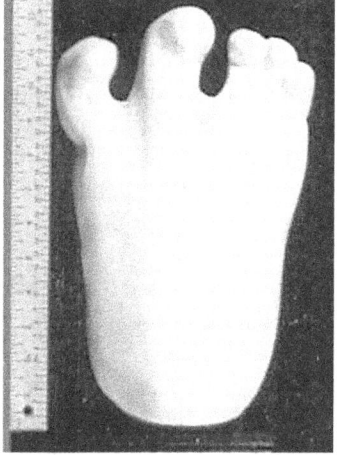

Yeti footprint cast.

An alleged yeti scalp (one of three known to exist) was professionally examined and declared to have been made from the skin of a serow, a member of the goat antelope family. It has been concluded that all known scalps are therefore *likely* fabrications.

It is possible, of course, that the scalp examined and the other two had been copied from an *original* yeti scalp. The yeti is held sacred in Tibet, so when one monastery *possibly* obtained a real scalp, all other monasteries wanted one. The monks in the other monasteries therefore made duplicate scalps.

Over the centuries all scalps would became "real" in the eyes and hearts of the monks. The inference here, therefore, is that hidden away in some lofty secluded monastery rests a real yeti scalp. We might just wonder if the monks who have the original would even allow it to be viewed by outsiders, let alone be taken away for analysis. Is it possible the researchers were sidetracked? Moreover, it is even possible that one of the two scalps known to exist but not examined is the real scalp. These scalps are about 350 years old. I do not have the specific age of the scalp examined.

Further, an alleged yeti skeletal hand held together with wire was also found and professionally examined. It was declared to be made from part human and part animal bones. However it is now known that a researcher had previously stolen some of the bones from the hand and replaced them with human bones. When the hand was examined, human bones were naturally determined. No specific identification was provided for the "animal" bones. Certainly, another examination should be performed on the hand, but I wonder if we could now get access to it – if it is still there. I have been informed that the entire hand was stolen in the late 1980s.

The latest information on the yeti appeared in *The Times* (Britain) on February 4, 2001. The entire story was later provided in a television documentary (*To the Ends of the Earth* series). We learn that a team of British scientist went on an expedition to Bhutan to seek evidence of the yeti's existence. Here they obtained the services of a resident "official yeti hunter." The yeti hunter told the scientist he had seen the creature enter a hollow at the base of a large cedar tree. He then led the scientists on a long arduous trek to the tree, which was situated in a forest in eastern Bhutan.

One of the scientists, Dr. Rob McCall, a zoologist, obtained hair strands from the entrance to the hollow. It appears the creature scraped its shoulders or upper back against the tree as it bent over to enter the hollow, thereby leaving hair strands.

The hair was analyzed in Britain by Bryan Sykes, Professor of Human Genetics at the Oxford Institute of Molecular Medicine. Sykes stated in the news release:

We found some DNA on it, but we don't know what it

is. It's not human, not a bear nor anything else we have so far been able to identify. It's a mystery and I never thought this would end in a mystery. We have never encountered DNA that we couldn't recognize before.

Nothing more has come to my attention, but I am sure further research is underway.

Like the sasquatch, the yeti has been given postage stamp distinction.

Top: Bhutan, 1966. Five views that were shown on 15 stamps of different denominations.

Center, left: Bhutan, 1970.

Center, right: Bhutan, 1996.

Bottom: Maldives Islands souvenir sheet, 1993. The stamp image shows a footprint discovered by Eric Shipton and Michael Ward in 1951. The information shown reads:

The Yeti: Giant footprints have been encountered in the Himalayan mountain snows since 1887. Sometimes 18 inches in length and 7 inches wide, the tracks have been attributed to the Yeti or Abominable Snowman.

APPENDIX

OHIO BIGFOOT INCIDENTS – COUNTY ORDER

COUNTY/DATE	LOCATION	TYPE OF INCIDENT/WITNESS
ADAMS		
1897/05/26	ROME	SIGHTING (C. LUKINS/BOB FORNER)
1995/05/00	FOREST AREA	SIGHTING, SOUNDS, ODORS (HUNTER)
1996/07/00	FOREST AREA	SIGHTING (RETIRED POLICE OFFICER)
1997/06/26	DUNKINSVILLE	SIGHTING, SOUNDS (RESIDENT)
2002/02/23	BENTONVILLE	SIGHTING, SNDS, F/H PTS SMITH/GREER)
ALLEN		
1950/00/00	BLUFFTON	SOUNDS, DOG KILLS (YOAKAMS)
1952/00/00	BLUFFTON	SOUNDS, DOG KILLS (YOAKAM BOY)
1956/11/29	BLUFFTON	SOUNDS/DOG KILLINGS (WATTS/ HTRS.)
1980/10/00	LIMA	SIGHTING (RESIDENT)
1980/00/00	LIMA	FECES (INVESTIGATOR)
1998/00/00	WESTMINSTER	KNOCKING SOUNDS/SIGHTING (4 MEN)
ASHLAND		
1997/07/03	MOHICAN RIVER	SIGHTING, ODOR, SOUNDS (STUDENTS)
ASHTABULA		
1954/00/00	ANDOVER	SIGHTING (DEAN AVERICK)
1974-1979	ASHTABULA	SIGHTING (REFERENCE)
1980/06/00	ROME	SIGHTING, SOUNDS, OTHER (FAMILY)
1981/06/00	ROME	SIGHTING,(REFERENCE)
1981/07/00	ROME	SIGHTING, FOOTPRINTS, ODORS (REF.
1981/07/00	ROME	SIGHTING, FOOTPRINTS, ODORS (REF.
AUGLAISE		
1979/00/00	WAPAKONETA	SIGHTING, FOOTPRINTS (MR. SHEETS)
BELMONT		
1991/10/26	ST. CLAIRSVILLE	SIGHTING (REFERENCE)
1995/00/00	FLUSHING	SIGHTING (REFERENCE - WHITE BF)
1995/04/14	FLUSHING	FOOTPRINTS, OTHER (RESEARCHERS)
1995/07/06	FLUSHING	FOOTPRINTS, OTHER (RESEARCHERS)
BROWN		
1996/03/00	RURAL AREA	SIGHTING (FARM LADY)
1997/04/06	UTOPIA	SOUNDS (RESIDENT)
BUTLER		
1997/02/00	HAMILTON	FOOTPRINTS (JACKIE SMITH)
CARROLL		
1978/09/09	MINERVA	SIGHTING, SOUNDS (HENRY COLT)
CHAMPAIGN		
1979/07/00	URBANA	SIGHTING (RONALD CHAMBERLIN)
1980/03/00	WOODSTOCK	FOOTPRINTS (RESIDENT)
1980-1989	FOREST AREA	SIGHTING, SOUNDS (5 PEOPLE)
1981/11/11	RURAL AREA	SIGHTING, FOOTPRINTS (REFERENCE)
1985/09/00	WOODSTOCK	SIGHTING (MAN AND WIFE)
1987/08/00	WOODSTOCK	SIGHTING, HAIR (REV. LEE BIRT)
CLERMONT		
1968/00/00	POINT ISABLE	SIGHTING (REFERENCE)
1976/04/04	MILFORD	SIGHTING (REFERENCE)

CLINTON
1961/00/00 WILMINGTON SIGHTING, SOUNDS (REFERENCE)
COLUMBIANA
1980/07/00 LISBON SIGHTING,FT.PRNTS,SNDS.OTR. (REF.)
1991/08/00 COLUMBIANA SIGHTING (REFERENCE)
1991/09/00 COLUMBIANA SIGHTING (REFERENCE)
COSHOCTON
1988/01/01 RURAL AREA SIGHTING (REFERENCE)
1992/00/00 FOREST AREA SIGHTING (RETIRED MECHANIC)
2000/05/07 WARSAW FOOTPRINTS (RESIDENT FARMER)
CUYAHOGA
1968/04/00 CLEVELAND ZOO SIGHTING (JOHN KEEL)
1972/08/00 CLEVELAND Z00 SIGHTING (WAYNE E. LEWIS)
DARKE
1981/08/00 RURAL AREA SIGHTING, ODORS (RESIDENTS)
ERIE
1970/10/00 HURON SIGHTING (MOTORIST)
FAIRFIELD
1897/04/00 STOUTSVILLE WILD MAN (RESIDENTS)
1970-1980 PLEASANTVILLE SIGHTING (REFERENCE)
FRANKLIN
1972/11/00 DUBLIN SIGHTING, FOOTPRINTS (REFERENCE)
1973/10/00 DUBLIN – GOLF C. SIGHTING (SECURITY GUARDS)
1974/09/00 WESTERVILLE FOOTPRINTS (RESIDENTS)
1997/06/00 RURAL AREA SIGHTING (MOTORIST)
GALLIA
1869/01/00 GALLIPOLIS WILD MAN? (RESIDENTS)
1912/00/00 RURAL AREA SIGHTING (RESIDENT)
1969/01/23 GALLIPOLIS SIGHTING (REFERENCE)
GREENE
1986/00/00 XENIA SIGHTING (REFERENCE)
1989/00/00 XENIA SIGHTING (REFERENCE)
HAMILTON
1959/00/00 CINCINNATI SIGHTING (TRUCKER)
1975/00/00 FOREST AREA UNUSUAL DOME STRUCTURES
1981/00/00 COLERAIN SIGHTING, SOUNDS (MARRIED COUPLE)
1984/00/00 CINCY-LUNKIN AP SIGHTING, FOOTPRINT (REFERENCE)
1984/07/02 SHARONVILLE SIGHTING (MOTHER AND 3 CHILDREN)
1995/09/00 SEDAMSVILLE SIGHTING (RESIDENT)
1995/00/00 MADISONVILLE FOOTPRINTS (RESIDENT/OBRSG)
HOCKING
1897/00/00 LOGAN SOUNDS, OTHER (FARMERS)
JEFFERSON
1979/00/00 BLOOMINGDALE FOOTPRINTS (BEVERLY FLETCHER)
1988/00/00 SMITHFIELD SIGHTING (FATHER AND SON)
1997/00/00 WINTERSVILLE SIGHTING (BOY AND GRANDFATHER)
KNOX
1978/07/00 KNOX CO. SIGHTING (MOTORIST)
LAWRENCE
1940/00/00 HANGING ROCK SIGHTING (MOTORIST/WIFE/CHILD)
1972/11/00 RURAL AREA SIGHTING (CAB DRIVER)
2001/07/00 LAKE VESUVIOUS SIGHTING, FOOTPRINTS (JAMES GOOD)

LOGAN
1979/10/07 (SIMPSON)	NORTH LEWISB.	SIGHTING, FOOTPRNTS,SNDS.
1980/00/00	NORTH LEWISB.	SIGHTING (RESIDENT FARMER)
1980/06/00	WEST MANSFIELD	SIGHTING (RESIDENT FARMER)
1980/06/00	WEST MANSFIELD	SIGHTING (RESIDENT FARMER)
1980/06/00	RUSSELL'S POINT	SIGHTING FOOTPRNTS, ODORS (FARMR)
1980/06/22	RURAL AREA	SIGHTING (RESIDENTS)
1980/06/25	RURAL AREA	SIGHTING (TOM QUAY)
1980/06/26	RURAL AREA	SIGHTING (LARRY RAMEY)
1981/00/00	NORTH LEWISB.	SIGHTING (REFERENCE)
1981/00/00	BELLEFONTAINE	SIGHTING (DEPUTY)
1981/00/00	BELLEFONTAINE	SIGHTING, FOOTPRINTS (MAMMAL R.T.)
1995/03/23	NORTH LEWISB.	FECES, OTHER (RESEARCHERS)

LORAIN
1973/08/00	OBERLIN	SIGHTING, FOOTPRNTS RANDOLPH/OTR.

MADDISON
1979/00/00	LONDON	SIGHTING (RESIDENTS)
1980-1990	LONDON	SIGHTING (RESIDENTS)
1980-1990	LONDON	FOOTPRINTS (RESIDENT/RESEARCHERS)
1980-1990	LONDON	SIGHTING (RESIDENT (MOTHER))
1980-1990	WEST JEFFERSON	SIGHTING, FOOTPRINTS (RESIDENT)
1980/07/15	PLAIN CITY	SIGHTING (C.LOVEJOY/R.WINN/FRNDS)
1981/00/00	LONDON	SIGHTING (RESIDENTS (CAR INCIDENT)
1981/07/00	LONDON	SIGHTING (REFERENCE)
1981/07/16	WEST JEFFERSON	SIGHTING (REFERENCE)
1985/00/00	LONDON	SIGHTING, FOOTPRTS,SNDS.(BOYS/10-14)
1985/09/00	WEST JEFFERSON	SIGHTING (BETTY POWELL)

MUSKINGUM
1897/06/00	HOPEWELL	WILD MAN? (MRS. BARTLEBAUGH)

PIKE
1999/12/22	LATHAM	SIGHTING (MAN AND GIRLFRIEND)

PORTAGE
1977/03/08	NELSON	SIGHTING (BARBARA PISTILLI)

PREBLE
1975/07/00	RURAL AREA	SIGHTING (THREE CHILDREN (10-12))
1976/00/00	EATON	SIGHTING, FOOTPRINTS (REFERENCE)
1977/05/18	RBTS COV. BRDGE	SIGHTING (TWO 13 YEAR-OLD BOYS)
1977/05/24	RURAL AREA	FOOTPRINTS (FARMER)
1977/12/00	EATON	SIGHTING, FOOTPRTS, ODORS (REF.)
1996/04/06	RURAL AREA	FOOTPRINTS, OTHER (RESEARCHERS)
1996/08/16	RURAL AREA	SIGHTING, ODORS (MOTHER AND DTR.)
1996/08/28	RURAL AREA	ODORS, OTHER (RESEARCHERS)

PUTNAM
1954/00/00	PANDORA	SIGHTING (REFERENCE)

RICHLAND
1959/03/00	MANSFIELD	SIGHTING (RESIDENTS)
1963/03/00	MANSFIELD	SIGHTING (REFERENCE)
1973/08/00	MANSFIELD	SIGHTING (FARMER)
1978/06/00	BUTLER	SIGHTING (RESIDENTS)
1978/07/08	BUTLER	SIGHTING (E. & K. KLINE - CHILDREN)
1978/07/12	BUTLER	SIGHTING, SDS, ODORS (TERESA KLINE)
1988/00/00	FOREST AREA	SOUNDS, FT. PR.SIGHTNG (TRAPPERS)

SCIOTO
1930/00/00	SCIOTOVILLE	SIGHTING (GEORGE JOHNSON (10)
2003/06/18	SHAWNEE ST PK.	FOOTPRINTS (MAN AND WIFE)

STARK
1957/00/00	ALLIANCE	SIGHTING (REFERENCE)
1973/08/00	MASSILON	SIGHTING, ODORS (RESIDENTS)
1973/10/00	MASSILON	SIGHTING, ODORS (RESIDENTS)
1978/00/00	PARIS	SIGHTING (MRS. CAYTON)
1978/00/00	PARIS	SIGHTING (FOUR RESIDENTS)
1978/08/21	PARIS	SIGHTING, FOOTPRINTS (CAYTONS/FRN)
1978/08/22	PARIS	SIGHTING (MARY ACKERMAN)
1978/08/23	PARIS	SIGHTING (HOWE CAYTON)
1978/08/26	PARIS	SIGHTING, FOOTPRTS, SNDS (NUTTERS)
1978/09/08	PARIS	SIGHTING (MARY ACKERMAN)
1979/00/00	PARIS	SIGHTING (HERBERT BURKE JR.)
1991/12/00	ALLIANCE	SIGHTING (RON BRUNNER)
1992/11/00	FOREST AREA	SIGHTING (BOW-HUNTER)

SUMMIT
1975/02/19	AKRON	NEST/DOME STRUCT.. (RESEARCHERS)
1979/01/09	CUYAHOGA FLS.	SIGHTING (REFERENCE)
1988/00/00	AKRON	SIGHTING (REFERENCE)
1985-1995	BARBERTON	IGHTING (BARBARA BILINOVICH)
1995/02/19	AKRON	POSSIBLE NEST, FOOTPRNTS (INVEST.)

TRUMBULL
1997/01/00	HUBBARD	SIGHTING, FOOTPR. (SCHOOL TEACHER)

TUSCARAWAS
1992/08/00	NEWCOMERSTN.	SIGHTING (REFERENCE)
1995/04/00	NEWCOMERSTN.	SIGHTING (REFERENCE)

UNION
1980/00/00	MARYSVILLE	SIGHTING (REFERENCE)
1980/06/17	RURAL AREA	SIGHTING, FOOTPRINTS (P. POLING)
1980/06/24	MARYSVILLE	SIGHTING (DONNA RIEGLER)
1985/06/19	RURAL AREA	SIGHTING CLAUDIA BEESON)

VAN WERT
1981/00/00	OHIO CITY	FOOTPRINTS, OTHER (RESIDENT)

VINTON
1970-1979	RURAL AREA.	SIGHTING (REFERENCE)
1980/00/00	MCARTHUR PK.	SIGHTING/SHOOTING/DAMAGE (YOUTH)
1980/08/24	FOREST AREA	SIGHTING (LARRY E. COTTRILL)
1980/10/11	MCARTHUR	SIGHTING (RODNEY PEOPLES)

SALT FORK STATE PARK
2004/08/18	FOREST AREA	SIGHTING, SOUNDS (COUPLE)

WAYNE NATIONAL FOREST
1966/00/00	FOREST AREA	SIGHTING (REFERENCE)
1980/08/00	FOREST AREA	SIGHTING FOOTPRINTS (VISITOR)
1980/10/00	FOREST AREA	FOOTPRINTS OTR (GARDINER/.COTRILL)
1994/00/00	FOREST AREA	FOOTPRINTS (FOREST RANGER)
1995/00/00	FOREST AREA	FOOTPRINTS (FOREST RANGE)

OHIO BIGFOOT INCIDENTS – DATE ORDER

DATE	LOCATION	TYPE OF INCIDENT/WITNESS
1869/01/00	GALLIA	WILD MAN ? (RESIDENTS)
1897/00/00	HOCKING	SOUNDS, OTHER (FARMERS)
1897/04/00	FAIRFIELD	WILD MAN (RESIDENTS)
1897/05/26	ADAMS	SIGHTING (C. LUKINS/BOB FORNER)
1897/06/00	MUSKINGUM	WILD MAN? (MRS. BARTLEBAUGH)
1912/00/00	GALLIA	SIGHTING (RESIDENT)
1930/00/00	SCIOTO	SIGHTING (GEORGE JOHNSON - 10)
1940/00/00	LAWENCE	SIGHTING (MOTORIST/WIFE/CHILD)
1950/00/00	ALLEN	SOUNDS, DOG KILL (YOAKAMS)
1952/00/00	ALLEN	SOUNDS, DOG KILL (YOAKAM BOY/FR.)
1954/00/00	ASHTABULA	SIGHTING (DEAN AVERICK)
1954/00/00	PUTNAM	SIGHTING (REFERENCE)
1956/11/29	ALLEN	SOUNDS/DOG KILL (WATTS/COON HTRS)
1957/00/00	STARK	SIGHTING (REFERENCE)
1959/00/00	HAMILTON	SIGHTING (TRUCKER)
1959/03/00	RICHLAND	SIGHTING (RESIDENTS)
1961/00/00	CLINTON	SIGHTING SOUNDS (REFERENCE)
1963/03/00	RICHLAND	SIGHTING (REFERENCE)
1966/00/00	WAYNE N.F..	SIGHTING (REFERENCE)
1968/00/00	CLERMONT	SIGHTING (REFERENCE)
1968/04/00	CUYAHOGA	SIGHTING (JOHN KEEL)
1969/01/23	GALLIA	SIGHTING (REFERENCE)
1970/10/00	ERIE	SIGHTING (MOTORIST)
1970-1979	VINTON	SIGHTING (REFERENCE)
1970-1980	FAIRFIELD	SIGHTING (REFERENCE)
1972/08/00	CUYAHOGA	SIGHTING (WAYNE E. LEWIS)
1972/11/00	FRANKLIN	SIGHTING, FOOTPRINTS (REFERENCE)
1972/11/00	LAWRENCE	SIGHTING (CAB DRIVER)
1973/08/00	RICHLAND	SIGHTING (FARMER)
1973/08/00	STARK	SIGHTING ODORS (RESIDENTS)
1973/08/00	LORAIN	SIGHTING FOOTPR. (R.RANDOLPH/OTRS)
1973/10/00	FRANKLIN	SIGHTING (SECURITY GUARDS)
1973/10/00	STARK	SIGHTING, ODORS, (RESIDENTS)
1974/09/00	FRANKLIN	FOOTPRINTS (RESIDENTS)
1974-1979	ASHTABULA	SIGHTING (REFERENCE)
1975/00/00	HAMILTON	UNUSUAL DOME STRUCTURES
1975/07/00	PREBLE	SIGHTING (3 CHILDREN (10-12)
1976/00/00	PREBLE	SIGHTING FOOTPRINTS (REFERENCE)
1976/04/04	CLERMONT	SIGHTING (REFERENCE)
1977/03/08	PORTAGE	SIGHTING (BARBARA PISTILLI)
1977/05/18	PREBLE	SIGHTING (TWO 13 YEAR-OLD BOYS)
1977/05/24	PREBLE	FOOTPRINTS (FARMER)
1977/12/00	PREBLE	SIGHTING FOOTPRTS, ODORS (REF.)
1978/00/00	STARK	SIGHTING (FOUR RESIDENTS)
1978/00/00	STARK	SIGHTING (MRS. CAYTON)
1978/06/00	RICHLAND	SIGHTING (RESIDENTS)
1978/07/00	KNOX	SIGHTING (MOTORIST)

1978/07/08	RICHLAND	SIGHTING (E. & K. KLINE - CHILDREN)
1978/07/12	RICHLAND	SIGHTING, SNDS, ODORS (KLINE -CHILD)
1978/08/21	STARK	IGHTING, FOOTPR.(CAYTONS/4 FRIENDS)
1978/08/22	STARK	SIGHTING (MARY ACKERMAN)
1978/08/23	STARK	SIGHTING (HOWE CAYTON)
1978/08/26	STARK	SIGHTING, FOOTPRTS, SNDS. (NUTTERS)
1978/09/08	STARK	SIGHTING (MARY ACKERMAN)
1978/09/09	CARROLL	SIGHTING, SOUNDS (HENRY COLT)
1979/00/00	JEFFERSON	FOOTPRINTS (BEVERLY FLETCHER)
1979/00/00	MADDISON	SIGHTING (RESIDENTS)
1979/00/00	STARK	SIGHTING (HERBERT BURKE JR.)
1979/00/00	AUGALIZE	SIGHTING, FOOTPRINTS (MR. SHEETS)
1979/01/09	SUMMIT	SIGHTING (REFERENCE)
1979/07/00	CHAMPAIGN	SIGHTING (RONALD CHAMBERLIN)
1979/10/07	LOGAN	SIGHTING FOOTPR.,SNDS.(S. SIMPSON)
1980/00/00	ALLEN	FECES (INVESTIGATOR)
1980/00/00	UNION	SIGHTING (REFERENCE)
1980/00/00	VINTON	SIGHTING/SHOOTING/DAMAGE (YOUTH)
1980/00/00	LOGAN	SIGHTING (RESIDENT FARMER)
1980/03/00	CHAMPAIGN	FOOTPRINTS (RESIDENT)
1980/06/00	ASHTABULA	SIGHTING, SOUNDS, OTHER (FAMILY)
1980/06/00	LOGAN	SIGHTING, FOOTPR., ODORS (FARMER)
1980/06/00	LOGAN	SIGHTING (RESIDENT FARMER)
1980/06/00	LOGAN	SIGHTING (RESIDENT FARMER)
1980/06/17	UNION	SIGHTING, FOOTPRINTS (P. POLING)
1980/06/22	LOGAN	SIGHTING (RESIDENTS)
1980/06/24	UNION	SIGHTING (DONNA RIEGLER)
1980/06/25	LOGAN	SIGHTING (TOM QUAY)
1980/06/26	LOGAN CO.	SIGHTING (LARRY RAMEY)
1980/07/00	COLUMBIANA	SIGHTING,FT.PRNTS, SNDS.OTR (REF.)
1980/07/15	MADDISON	SIGHTING (C.LOVEJOY/R.WINN/FRNDS)
1980/08/00	WAYNE N.F.	SIGHTING, FOOTPRINTS (VISITOR)
1980/08/24	VINTON	SIGHTING (LARRY E. COTTRILL)
1980/10/00	WAYNE N.F.	FOOTPR.,OTHR (GARDINER/.COTRILL)
1980/10/00	ALLEN	SIGHTING (RESIDENT)
1980/10/11	VINTON	SIGHTING (RODNEY PEOPLES)
1980-1989	CHAMPAIGN	SIGHTING, SOUNDS (5 PEOPLE)
1980-1990	MADDISON	FOOTPRINTS (RESIDENT/RESEARCHERS)
1980-1990	MADDISON	SIGHTING (RESIDENT (MOTHER))
1980-1990	MADDISON	SIGHTING (RESIDENTS)
1980-1990	MADDISON	SIGHTING, FOOTPRINTS (RESIDENT)
1981/00/00	LOGAN	SIGHTING (DEPUTY)
1981/00/00	LOGAN	SIGHTING, FOOTPR. (MAM. RES. TEAM)
1981/00/00	HAMILTON	SIGHTING, SOUNDS (MARRIED COUPLE)
1981/00/00	MADDISON	SIGHTING (RESIDENTS (CAR INCIDENT))
1981/00/00	LOGAN	SIGHTING (REFERENCE)
1981/00/00	VAN WERT	FOOTPRINTS, OTHER (RESIDENT)
1981/06/00	ASHTABULA	SIGHTING (REFERENCE)
1981/07/00	MADDISON	SIGHTING (REFERENCE)
1981/07/00	ASHTABULA	SIGHTING, FOOTPRINTS, ODORS (REF.)
1981/07/00	ASHTABULA	SIGHTING, FOOTPRINTS, ODORS (REF.)
1981/07/16	MADDISON	SIGHTING (REFERENCE)

Date	County	Description
1981/08/00	DARKE	SIGHTING, ODORS (RESIDENTS)
1981/11/11	CHAMPAIGN	SIGHTING, FOOTPRINTS (REFERENCE)
1984/00/00	HAMILTON	SIGHTING, FOOTPRINT (REFERENCE)
1984/07/02	HAMILTON	SIGHTING (MOTHER AND 3 CHILDREN)
1985/00/00	MADDISON	SIGHTING, FOOTPR.,SNDS. (BOYS 10-14)
1985/06/19	UNION	SIGHTING (CLAUDIA BEESON)
1985/09/00	MADDISON	SIGHTING (BETTY POWELL)
1985/09/00	CHAMPAIGN	SIGHTING (MAN AND WIFE)
1985-1995	SUMMIT	SIGHTING (BARBARA BILINOVICH)
1986/00/00	GREENE	SIGHTING (REFERENCE)
1987/08/00	CHAMPAIGN	SIGHTING, HAIR (REV. LEE BIRT)
1988/00/00	SUMMIT	SIGHTING (REFERENCE)
1988/00/00	RICHLAND	SOUNDS, FT. PR. SIGHTING (TRAPPERS)
1988/00/00	JEFFERSON	SIGHTING (FATHER AND SON)
1988/01/01	COSHOCTON	SIGHTING (REFERENCE)
1989/00/00	GREENE	SIGHTING (REFERENCE)
1991/08/00	COLUMBIANA	SIGHTING (REFERENCE)
1991/09/00	COLUMBIANA	SIGHTING (REFERENCE)
1991/10/26	BELMONT	SIGHTING (REFERENCE)
1991/12/00	STARK	SIGHTING (RON BRUNNER)
1992/00/00	COSHOCTON	SIGHTING (RETIRED MECHANIC)
1992/08/00	TUSCARAWAS	SIGHTING (REFERENCE)
1992/11/00	STARK	SIGHTING (BOW-HUNTER)
1994/00/00	WAYNE N.F.	FOOTPRINTS (FOREST RANGER)
1995/00/00	BELMONT	SIGHTING (REFERENCE (WHITE BF))
1995/00/00	WAYNE N.F.	FOOTPRINTS (FOREST RANGER)
1995/00/00	HAMILTON	DONE STRUCTURES (RESEACHERS)
1995/00/00	HAMILTON	FOOTPRINTS (RESIDENT/OBRSG)
1995/02/19	SUMMIT	POSSIBLE NEST, FOOTPR. (RESEARCRS)
1995/03/23	LOGAN	FECES, OTHER (RESEARCHERS)
1995/04/00	TUSCARAWAS	SIGHTING (REFERENCE)
1995/04/14	BELMONT	FOOTPRINTS, OTHER (RESEARCHERS)
1995/05/00	ADAMS	SIGHTING, SOUNDS, ODORS (HUNTER)
1995/07/06	BELMONT	FOOTPRINTS, OTHER (RESEARCHERS)
1995/09/00	HAMILTON	SIGHTING (RESIDENT)
1996/03/00	BROWN	SIGHTING (FARM LADY)
1996/04/06	PREBLE	FOOTPRINTS, OTHER (RESEARCHERS)
1996/07/00	ADAMS	SIGHTING (RETIRED POLICE OFFICER)
1996/08/16	PREBLE	SIGHTING, ODORS (MOTHER/DAUGTR)
1996/08/28	PREBLE	ODORS, OTHER (INVESTIGATORS)
1997/00/00	JEFFERSON	SIGHTING (BOY AND GRANDFATHER)
1997/01/00	TRUMBULL	SIGHTING, FOOTPR. (SCHOOL TEACHER)
1997/02/00	BUTLER	FOOTPRINTS (JACKIE SMITH)
1997/04/06	BROWN	SOUNDS (RESIDENT)
1997/06/00	FRANKLIN	SIGHTING (MOTORIST)
1997/06/26	ADAMS	SIGHTING, SOUNDS (RESIDENT)
1997/07/03	ASHLAND	SIGHTING, ODOR, SOUNDS (STUDENTS)
1998/00/00	ALLEN	KNOCKING SOUNDS/SIGHTING (4 MEN)
1999/12/22	PIKE	SIGHTING (MAN AND GIRLFRIEND)
2000/05/07	COSHOCTON	FOOTPRINTS (RESIDENT FARMER)
2001/07/00	LAWRENCE	SIGHTING, FOOTPRINTS (JAMES GOOD)
2002/02/23	ADAMS	SIGHTING, SNDS, F/H PR. (SMITH/GREER)
2003/06/18	SCOTIO	FOOTPRINTS (MAN AND WIFE)
2004/08/18	SALT FORK SP	SIGHTING/SNDS (COUPLE)

OHIO'S COUNTIES

Adams (C11), Allen (B4), Ashland (E4), Ashtabula (H1), Athens (F9), Auglaize (B5), Belmont (H7), Brown (B11), Butler (A9), Carroll (H5), Champaign (C6), Clark (C7), Clermont (B10), Clinton (B9), Columbiana (H4), Coshocton (F6), Crawford (E4), Cuyahoga (G2), Darke (A6), Defiance (A2), Delaware (D6), Erie (E2), Fairfield (E8), Fayette (C8), Franklin (D7), Fulton (B1), Gallia (E11), Geauga (H2), Greene (B8), Guernsey (G7), Hamilton (A10), Hancock (C3), Hardin (C5), Harrison (H6), Henry (B2), Highland (C10), Hocking (E9), Holmes (F5), Huron (E3), Jackson (E10), Jefferson (H6), Knox (E5), Lake (H1), Lawrence (E12), Licking (E7), Logan (C5), Lorain (F2), Lucas (C1), Madison (C7), Mahoning (H3), Marion (D5), Medina (F3), Meigs (F10), Mercer (A5), Miami (B7), Monroe (H8), Montgomery (B8), Morgan (F8), Morrow (E5), Muskingum (F7), Noble (G8), Ottawa (D1), Paulding (A3), Perry (F8), Pickaway (D8), Pike (D10), Portage (H3), Preble (A8), Putnam (B3), Richland (E4), Ross (D9), Sandusky (D2), Scioto (D11), Seneca (D3), Shelby (B6), Stark (H4), Summit (G3), Trumbull (H3), Tuscarawas (G5), Union (C6), Van Wert (A4), Vinton (E10), Warren (B9), Washington (G9), Wayne (F4), Williams(A1), Wood (C2), Wyandot (C4).

141

Bibliography

Books

Bayanov, Dmitri. *1996. In the Footsteps of the Russian Snowman.* Crypto-Logos Publishers, Moscow, Russia.
Bondeson, Jan. 1997. *A Cabinet of Medical Curiosities.* Cornell University Press, New York, NY, U.S.A.
Encyclopedia Canadiana. 1970.
Napier, John. 1972. *Bigfoot – Startling Evidence of Another Form of Life on Earth Now.* Berkley Publishing, New York, NY, U.S.A.
Patterson, Roger. 1966. *Do Abominable Snowmen of America Really Exist?* Franklin Press, Yakima, Washington, U.S.A.
Roosevelt, Theodore. 1892. *Wilderness Hunter, Outdoor Pastimes of an American Hunter.* G.P. Putnan's Sons, New York, NY, U.S.A
Washington Environmental Atlas. 1975. U.S. Army Corps of Engineers

Newspapers and Magazines
(Book page numbers are referenced)

Akron Beacon Journal, Ohio (80, 83, 96)
Akron, Ohio Beacon & Republican – Buckeye News (61)
Associated Press (14)
Bluffton News, Ohio (46)
Chronicle Telegram, The, Ohio (65)
Cincinnati Post and Times Star, Ohio (58)
Cleveland Magazine, Ohio (67)
Cleveland Plain Dealer, The, Ohio (40, 53, 54, 63, 71, 81, 91)
Cleveland Press, Ohio (92)
Columbus Citizen Journal, Ohio, (93, 94)
Columbus Dispatch, Ohio, (69, 76)
Daily Olympian, Washington, (93)
Enquirer, The, (early newspaper), Ohio (55)
Forest Press, Ohio (49)
Homestead City News, Ohio (48)
Humboldt Times, California (14)
Jackson Journal Herald, Ohio (51,70)
Journal Herald, Dayton, Ohio (53)
Minnesota Weekly Record (56)
Mount Vernon News, Ohio (63)
National Enquirer, The (25)
New York Globe, The (7, 65, 91)

News Register, West Virginia (61)
Newsletter of Intriguing Studies, Ohio (80)
Probe the Unknown (magazine) California (15)
Record Courier, Ohio (72)
San Francisco Chronicle (7)
Times, The, London, England (132)
True magazine (111)
Wall Street Journal, New York (83)
Wapakoneta, Ohio News (47, 66)
Washington Star News (25, 29)
World Journal, Taiwan (129)

Photographs / Drawings / Illustrations
Sources and Copyrights

PAGE	DESCRIPTION	SOURCE/COPYRIGHT
F. COVER	Sasquatch head	C.L. Murphy
2	Clappison and Cook	J. Cook
8	Map	C. Murphy/Y. Leclerc
10	Map	Author's collection
12	Map	C.L Murphy/Y. Leclerc
13	Gigantopithecus and Krantz	Dian Horton
13	Three skulls	C.L. Murphy
13	Gigantopithecus – front view	C.L. Murphy
13	Gigantopithecus – side view	C.L. Murphy
14	Cast – Jerry Crew	C.L. Murphy
16	Mummified remains	Public Domain
19	Native mask – top	P. Travers
19	Native mask – lower	P. Travers
20	Native mask – side view	C.L. Murphy
20	Native mask – front view	C.L. Murphy
21	Delaware warning post	J. Cook
21	Delaware ax	J. Cook
26	*Washington Env. Atlas* page	Public Domain
30	Map – Washington	Author's collection
32	Sasquatch stamp – left	Canada Post
32	Sasquatch stamp – right	C.L. Murphy
33	Sasquatch study – 4 photos	C.L. Murphy/E&M Dahinden
34	Sasquatch profile – left	C.L.Murphy/E&M Dahinden
34	Sasquatch profile – right	C.L. Murphy/Y. Leclerc
34	Sasquatch foot/human foot	C.L. Murphy
34	Sasquatch foot – top view	C.L. Murphy
34	Sasquatch foot – front view	C.L. Murphy
35	Bourtsev statue – 4 views	C.L. Murphy

36	Birnam heads – 4 photos	C.L. Murphy
37	Murphy Sasq. bust – 3 views	C.L. Murphy
38	Patterson sasquatch bust	C.L. Murphy
40	Hand print	J. Cook
41	Sasquatch jumping up bank - drawing	J. Cook
41	Diagram of earth bank	J. Cook
42	Diagram of hand on branch	J. Cook
43	Hand cast – four fingers	J. Cook
43	Hand cast - index finger	J. Cook
44	Hand cast – photo/diagram	C.L. Murphy
44	Joedy Cook with hand cast	J. Cook
45	Ohio hand cast illustration	J. Cook/Yvon Leclerc
52	Footprint and shoe print	O. C. Bigfoot Studies
54	Footprint and ruler	D. Keating
59	Sasquatch drawing	O.C. Bigfoot Studies
52	Footprints in a series	O.C. Bigfoot Studies
52	Footprint and ruler	O.C. Bigfoot Studies
61	Footprint in snow	O.C. Bigfoot Studies
63	Footprint and ruler	J. Good
67	Footprint – top	J. Cook/G. Clappison
67	Footprint with hand pointing	J. Cook/G. Clappison
71	Footprint – top	G. Clappison
71	Footprint with hand – lower	G. Clappison
80	Cast – J. Cook	C.L. Murphy
87	Nest in trees	J. Cook/G. Clappison
87	Nest in trees close-up	J. Cook/G. Clappison
88	Summit dome /G.Clappison	J. Cook/G. Clappison
89	Hamilton dome/G.Clappison (top)	J. Cook/G. Clappison
89	Hamilton dome/G.Clappison (lower)	J. Cook/G. Clappison
90	Footprint with ruler	O.C. Bigfoot Studies
90	Cast – California (R. Titmus, 1958)	C.L. Murphy
92	Casts – Footprints and hand print	O.C. Bigfoot Studies
93	Tree carvings	R. Morgan
94	Casts – Wayne National Forest	O.C. Bigfoot Studies
94	Cast Comp. – L. cast	O.C. Bigfoot Studies
94	Cast Comp.– R. cast (P. Byrne, 1960)	C.L. Murphy
100	Cast – Double-tracked bear prints	Dian Horton
100	Cast – California (R. Titmus, 1958)	C.L.Murphy
101	Casts - D.T. bear prints/bigfoot	C.L.Murphy
101	Foot – P/G film and bear foot inset	C.Murphy/Dahindens
102	Matt Moneymaker	M. Moneymaker
112	Cast – California (R. Patterson, 1964)	C.L. Murphy
113	Film site model	C.L. Murphy
113	Frame 352 – P/G film	Dahindens
114	Casts – P/G film site (Patterson, 1967)	C.L. Murphy
116	Patterson and Gimlin	Dahindens
119	Cast/Heel, Skookum (R. Noll), 2000	C.L. Murphy

120		Cast Calif./dermals (Green, 1967)	C.L. Murphy
121		Krao with her circus manager	Public Domain
122		Julia Pastrana	Public Domain
124		Zana taunted by children - drawing	Lydia Bourtseva
125		Zana holding first newborn - drawing	B. Bannon
126		Igor Bourtsev holding Khwit's skull	I. Bourtsev
126		Khwit	I. Bourtsev
127		Karapetian Hominid - drawing	Unknown source
128		Karapetian	V. Karapetian
128		Bassou	B. Heuvelmans
129		Chinese article	*World Journal*
130		Francis de Loys' man-beast	F. de Loys (Pub.Dom.)
131		Cast – yeti footprint	C.L. Murphy
133		Stamps – Bhutan (6 stamps)	Bhutan Postal Service
133		Stamp – souvenir sheet, Maldives Is.	Maldives Islands P.S.
141		Map of Ohio showing counties	Author's collection
Back cover		C.L. Murphy	C.L. Murphy
Back cover		Joedy Cook	C.L. Murphy
Back cover		George Clappison	George Clappison

General Index

A

A Buckeye Boyhood (book), 20
A Cabinet of Medical Curiosities, 122
Abbott, Don, 110
Able Detective and Security Systems, 56
Ackerman, Mary, Mrs., 82, 83
Adams County, 40
Adams, Ester, 24
Ahwahneechees, 17
Akron Beacon Journal, 80, 83, 96
Akron, Ohio Beacon & Republican – Buckeye News, 61
Akron, Ohio, 60, 84, 102
Allegheny Plateau, 11
Allen County, 46
Alliance, Ohio, 80, 83
Andover, Ohio, 48
Anthropoid Research and Evaluation Center, 49
Appalachian Plateau, 11
Army Corps of Engineers (U.S.), 25
Ashland County, 47, 63
Ashtabula City, Ohio, 48
Ashtabula County, 48
Associated Press, 14
Atkins, Dale (pseudonym), 84,85,86,87,96
Atkins, Tim (pseudonym), 84, 85, 86, 87, 96
Auglaize County, 49
Averick, Dean, 48
Ayuukhl Nisga'a Department, 22

B

Barberton, Ohio, 84
Bartlebaugh, Rev. G.A., (wife of), 71
Bassou, 128, 129
Bayanov, Dmitri, 124
Beach, J.W., 123
Beebe, Frank L, 110
Beeson, Claudia, 91
Bella Coola, British Columbia, 98, 110
Bellefontaine, Ohio, 65, 66
Bellville, Ohio, 79
Belmont County, 49
Bentonville, Ohio, 41
Bering Strait, 12
Berlin Lake, Ohio, 108, 109
Big Darby Creek, 68
Bigfoot – Startling Evidence of Another Form of Life on Earth Now, 110
Bigfoot Field Researchers Organization, 93
Bigfoot Research Center, China, 129
Bigfoot Research Project, The, 28
Bilinovich, Barbars, 84
Birnam, Penny, 36
Birt, Lee, Reverend, 53, 64
Bitterroot Mountains, 57
Bloomingdale, Ohio, 61,62
Blue Creek Mountain, California, 94
Blue Mountains, Washington, 45
Bluegrass Region, Ohio, 11
Bluff Creek, California, 7, 14, 38, 99, 111, 112, 116
Bluffton News, 46
Bluffton, Ohio, 46
Board of Commissioners of Skamania County, Washington, 30, 32
Bondeson, Jan, Dr., 122,123
BoneClones, 43,44
Bosjesman, 21

146

Bourtsev, Igor, 35, 125
Bowling Green, Ohio, 15
Brainerd, S. Professor, MD, 123
British Columbia Provincial Museum (Royal Museum), 110
Brown County, 51
Brunner, Ron, 83
Burke, Herbert, Jr., 83
Butler County, 51
Butler, Ohio, 76, 77, 78, 79
Byrne, Peter, 28, 94

C

Calder, Peter, 23
Cambridge, Ohio, 93
Canada Post, 32
Canada's Legendary Creatures (postage stamps), 32
Carroll County, 51, 102, 109
Cayton, Herbert, 80, 81, 82
Cayton, Howe, 81, 82
Cayton, Mrs., 70, 80, 81, 82, 83
Chamberlin, Ronald, 52
Champaign County, 52
Chehalis First Nations, 20
Chilcutt, Jimmy, 42
Christian County, Kentucky, 15
Chronicle Telegram, The, 65
Cincinnati Post and Times Star, 58
Cincinnati, Ohio, 58
Ciochon, Russell, 13
Clappison, George, 2, 29, 74, 75, 84, 88, 89

Clayton, Peter, 24
Clermont County, 53
Cleveland Magazine, 67
Cleveland Plain Dealer, The, 40, 53, 54, 63, 71, 81, 91
Cleveland Press, 92
Cleveland Zoo, 54
Clinton County, 53
Colerain Township, Ohio, 58
Colt, Henry, 51
Columbiana City, Ohio, 53
Columbiana County, 53, 109
Columbus Citizen Journal, 93, 94
Columbus Dispatch, 69, 76
Columbus Zoo, 91
Cook, Joedy, 2, 28, 29, 41, 42, 44, 80, 84
Coppei Falls, Washington, 109
Coshocton County, 53, 107
Coshocton, Ohio, 18
Cottrill, Dean, 94
Cottrill, Larry E., 92
Coverington, Kentucky, 58
Creeger, Henry, 71
Crew, Gerald (Jerry), 14, 111, 112
Cuyahoga County, 54
Cuyahoga Falls, Ohio, 84

D

Daily Olympian, 93
Daniel Boone National Forest, Kentucky, 19
Darke County, 55
Darwin, Charles, 123
De Loys, Francis, 130

Dean's Boat Landing, 48
Delaware First Nations, 21
DeWerth, Marc, 93
Do Abominable Snowmen of America Really Exist?, 112
Dublin, Ohio, 55
Dunkinsville, Ohio, 41

E

Eaton, Ohio, 72, 73, 74, 75
Encyclopedia Canadiana, 15
Endres, Terry, 84
Enquirer, The, 55
Erie County, 55

F

Fairfield County, 55
Fairlawn area (Akron, Ohio), 84
FBI, 25, 27,28, 29
Federal Bureau of Investigation, 25,27, 28,29
Fletcher, Beverly, 61
Flushing, Ohio, 49, 50
Forest Press, The, 49
Forner, Bob, 40
Franklin County, 55
Franklin County Sheriff's Dept., 56
Freedom of Information – Privacy Act, 28
Freeman (Paul), 45
Friendship Lake, Ohio, 62
Frontz, Sergeant, 79

G

Gallia County, 56
Gallipolis, Ohio, 56, 58
Gardiner, Bob, 92, 94

Gigantopithecus blacki, 13
Gilford Pinchot National Forest, Washington, 119
Gilyuk, 71
Gimlin, Bob, 7, 38, 99, 111, 113, 114, 116
Ginluuak, 22
Glickman, Jeff, 114
Good, James, 63, 64
Grassman, 73, 86, 96
Graves, Pat, 112
Great Lakes Plains, Ohio, 11
Green, John, 97
Greene County, 58
Greer, Loren, 41
Greer, Tony, 4, 42
Guernsey County, 58, 104

H
Hair and Fiber Unit – FBI, 29
Hamilton County, 58, 89, 95
Hamilton, Ohio, 51
Hanging Rock Hill, Ohio, 63
Hanging Rock, Ohio, 63
Hauenstein, Wayne, 47
Hildreth, S. P., Dr., 18
Hocking County, 60
Hoffman, Richard, 72
Homstead City News, 48
Hopewell Township, 70
Horn, Fred, Chief (police), 79
Hosier National Forest, Indiana, 19
Hubbard, Ohio, 89
Humboldt Times, 14
Huron, Ohio, 55

I
In the Footsteps of the

Russian Snowman, 124
Institute of Forensic Medicine, Rikshospitalet, Oslo, Norway, 123
Ironton, Ohio, 63

J
Jack Nicholson Golf Course, 55
Jackson Journal Herald, 51, 70
Jeffers, Joan, 109
Jefferson County, 61, 62
Johnson, George, 80
Journal Herald, Dayton, Ohio, 53

K
Karapetian, V.S., Lt. Col., MD, 127, 128
Keating, Don, 50, 51, 54
Keck, Mrs., 81
Keel, John, 54
Kenmore area, (Akron, Ohio), 84, 85, 86, 89, 95, 96
Khwit, 125,126
Killbuck River, Ohio, 18
Kline, Eugene,76, 77, 78, 79
Kline, Kathy, 76, 77, 78
Kline, Roger, 77, 78, 79
Kline, Teresa, 76, 78, 79
Knox County, 62, 63
Krantz, Grover S., Dr., 13, 120, 126
Krao, 121

L
Lafayette, Ohio, 46, 47
Lake Vesuvius, Ohio, 63
Latham, Ohio, 71
Lawrence County, 63
Leclerc, Yvon, 34, 45
Lent, Mr. (Manager for Julia Pastrana), 123

Lewis, Wayne E. 54
Lexington, Kentucky, 15
Lima, Ohio, 47
Lisbon, Ohio, 53
Logan County, 64
Logan, Ohio, 60
London, Ohio, 68, 69, 70
Londonville, Ohio, 47
Long Lake, Ohio, 87
Lorain County, 67
Louisville, Kentucky, 15
Lovejoy, Charles, 68
Lukins, Charles, 40

M
Madison County, 68
Madisonville, Ohio, 58
Mammal Research Team of Lima, The, 91
Mansfield, Ohio, 75, 76, 77, 78
Marirupa Falls, Venezuela, 131
Marysville, Ohio, 91
Mashkovtsev, (Alexander, Dr.), 124
Massilon, Ohio, 80
McArthur (town), 92, 93
McArthur State Park, Ohio, 92
McCall, Rob, Dr., 132
McManus, Jean, 25
Meldrum, Jeff, Dr., 43
Michigan Anomaly Research, 56
Milford, Ohio, 53
Mill Bay, British Columbia, 24
Minerva, 51,82
Minnesota Weekly Record, 56
Mississippi River, 18
Mohican River, Ohio, 47
Moneymaker, Dr., 104
Moneymaker, Matt, 102, 103, 104, 105, 107, 108, 109

Morgan, Robert, 93
Moscow, Russia, 123
Mott, Alexander B., Dr., 123
Mount Vernon News, 63
Mountain monkeys, 20
Munns, William, 13
Muskingum County, 70

N

Napier, John, Dr., 110
Nass Harbor, British Columbia, 22
Nass River, British Columbia, 23, 24
National Enquirer, The, 25
Nelson, Ohio, 71
New York Globe, The, 7, 65, 91
Newcomerstown, *Ohio*, 90
News Register, 61
Newsletter of Intriguing Studies, 80
Nisga'a Lisims Government, 22
Nisga's (native people), 19
North American Science Institute, 114
North Lewisburg, Ohio, 64, 65, 66
Nutter, Jerry, 82
Nutter, John, 82, 83
Nyce, Allison, 22
Nyce, Emma, 23, 24

O

O'Shanter, Tam, 21
Oberlin, Ohio, 67
Ogle's Woods, Ohio, 70
Ohio Bigfoot Research & Study Group, 2, 85, 96
Ohio Canal, 86

Ohio Center for Bigfoot Studies, 41, 80
Ohio City, Ohio, 91
Ohio County, Indiana, 69
Ohio Department of Natural Resources, 100
Ohio Mammal Research Team, 65
Ohio River, 11, 18, 58, 63
Ohio State University, 53, 61
Ohio Valley, 11, 12, 15, 19
Olmed (people), 15
Oslo, Norway, 123

P

Padanaram, Ohio, 48
Paducah, Kentucky, 18
Pamir Mountains (U.S.S.R/Russia), 27
Pandora, Ohio, 75
Para-Hominoid Research, Dispatch State Service (Columbus Dispatch), 76
Paris Township, Ohio, 80, 81, 83
Pastrana, Julia, 122, 123
Patterson, Roger, 7, 38, 99, 111, 112, 113, 114, 116
Patterson/Gimlin film, 20, 33, 35, 36, 101
Peoples, Rodney, 93
Pike County, 71
Pilichis, Dennis, 49
Pistilli, Barbara, Mrs., 71, 72
Plain City. Ohio, 68
Pleasantville, Ohio, 55
Point Isable, Ohio, 53

Point, Ambrose, 20
Poling, Patrick, 90
Porshnev, (Boris, Professor), 124, 125
Portage County, 71, 108
Powell, Betty, 70
Preble County Sheriff's Department, 72
Preble County, 72
Probe the Unknown (magazine), 15
Putnam County, 75

Q

Quay, Tom, 65

R

Ramey, Larry, 65
Randolph, Rudy, 67
Rates, M., 122
Record Courier, 72
Rice, Steve, Dr., 28
Richland County Sheriff's Office, 77, 78
Richland County, 62, 63, 75, 77, 78
Riegler, Donna, 91
Ringling Brothers, 121
Roberts Covered Bridge, 72
Roger's Auction (area), Ohio, 53
Rome, Ohio, 40, 49
Roosevelt, Theodore, 57
Russell's Point, Ohio, 65

S

Salt Fork State Park, Ohio, 93
Samuels, Gary, 131
San Francisco Chronicle, 7
Sanderson, Ivan T., 111
Sawvel, John, 96
Schaffner, Ron, 81
Scioto County, 80

Sciotoville, Ohio, 80
Shannon, Deputy, 83
Shawnee State Park, Ohio, 80
Sheets, Mr., 49
Shierley, Fred, Major, 26
Shipton, Eric, 133
Simpson farm (North Lewisburg), 64, 66, 67
Simpson, Mrs., 66
Simpson, Scott, 64
Skamania County, Washington, 30, 31, 32
Skookum cast, 119
Skookum Meadows, Washington, 119
Smedley, Richard, 15
Smith Dana, 41
Smithfield, Ohio, 62
St. Clairsville, Ohio, 49
Stark County, 70, 80, 102, 104, 109
Stevens, Albert, 22
Stevens, Horace, 22, 23
Stevens, William, 23
Stortz, Phil (acting police chief), 76
Stout, Ohio (now Stoutsville), 55
Stoutsville, Ohio, 55
Summit County, 60, 84
Sykes, Bryan, Professor, 132, 133
Sylvester, Roberty (deputy), 82

T

Terrace, British Columbia, 23
Till Plains, Ohio, 11
Tilman, Larry, 64
Tilman, Peggy, 64
Times, The, 132
To the Ends of the Earth (television series), 132
Titmus, Bob, 90, 100

Toltec (people), 15
True magazine, 111
Trumbull County, 89
Tsimshian (people), 19
Turolla, Pinto, 131
Tuscarawas County, 90

U

Union County Sheriff's Department, 91
Union County, 90
United States Department of the Interior, 19
Urbana, Ohio, 52
Utica, New York, 21
Utopia, Ohio, 51

V

Valley of Dades, Morocco, 128
Van Wert County, 91
Vancouver Museum (British Columbia), 22, 34
Venable, William, 20
Victoria, British Columbia, 110
Vinton County, 92

W

Wall Street Journal, 83
Wapakoneta, Ohio News, 47, 66
Wapakoneta, Ohio, 49
Ward, Michael, 133
Warsaw, Ohio, 54
Washington Environmental Atlas, 25, 28
Washington Star News, 25, 29
Watt, Richard, Mr. and Mrs., 47
Wayne National Forest, Ohio, 18, 64, 94

West Jefferson, Ohio, 68, 70
West Mansfield, Ohio, 64, 65
Westerville, Ohio, 56
Westminster, Ohio, 47
Whatcom County, Washington, 30
Wilderness Hunter, Outdoor Pastimes of an American Hunter, 57
Wills Creek Ohio, 107
Wilmington, Ohio, 53
Wilson, Pete, 51
Winn, Ron, 68
Wintersville, Ohio, 62
Woodstock, Ohio, 52
Wooly-Booger, 93
World Journal, 129

X

Xenia, Ohio, 58

Y

Yakima Reservation, Washington, 112
Yakima, Washington, 111
Yeti, 110, 131
Yoakam, Dallas, Mr. and/or Mrs., 46, 47
Yosemite Valley, California, 15, 16

Z

Zana, 124, 125

Raincoast Sasquatch
J. Robert Alley
0-88839-508-6
5½ x 8½, sc, 360 pages

Bigfoot Sasquatch Evidence
Dr. Grover S. Krantz
0-88839-447-0
5½ x 8½, sc, 348 pages

The Locals
Thom Powell
0-88839-552-3
5½ x 8½, sc, 271 pages

Sasquatch Bigfoot
Thomas N. Steenburg
0-88839-312-1
5½ x 8½, sc, 126 pages

All titles available at:
HANCOCK HOUSE PUBLISHERS
www.hancockhouse.com • sales@hancockhouse.com

Meet the Sasquatch
Christopher L. Murphy
0-88839-573-6
8½ x 11, sc
240 pages

The Bigfoot Film Controversy
Roger Patterson & Chris Murphy
0-88839-581-7
5½ x 8½, sc
264 pages

Sasquatch: *The Apes Among Us*
John Green
0-88839-123-4
5½ x 8½, sc
496 pages

All titles available at:
HANCOCK HOUSE PUBLISHERS
www.hancockhouse.com • sales@hancockhouse.com